Appreciations

It is important to me that I recognize some of my personal friends who have helped with this book by sharing their time and energy, and providing support.

I extend a very warm thanks to Dr. H. Frederick Vogt for his love and spiritual guidance in both my business and personal life. Those closest to us are our best teachers, and Dr. Fred is one of my finest teachers.

Kay Woodburn, close friend and former business partner, has also been a teacher. I have learned much from Kay in many areas. In the weight control programs that we organized, Kay communicated, through personal experience, her ideas—and people listened! Some of her research and methods, as applied to the overweight, are related directly in this book. Kay was instrumental in my growth and understanding regarding overweight people.

To my sister, Marilyn, who has provided a sounding board for me and also a testing ground for many of the ideas contained within: I love you. To a new-found friend, Elizabeth Scott Anderson, thank you. Your patience and guidance in editing this book were most needed. I guess I was just "lucky" to have found you. Finally, to Diane White, who supplied the main ingredient for writing this book, my warmest thoughts go out to you always.

Oh sister!
Love sprouts the wings of flight

your right to fly

Dr. James E. Melton

Global Publications
P.O. Box 2112 - Palm Springs, CA 92263

Library of Congress Catalog Card Number 80-82961
ISBN 0-9604752-0-6

Foreword

by C. Norman Shealy, MD, PhD

Many international medical scientists now believe that the next great advance in the health of Americans will occur through self-responsibility from using their own intelligence and initiative to achieve health. The primary requirement for such responsibility is education, and *Your Right to Fly* provides a good beginning for that approach. Of course, we have known for many centuries, perhaps thousands of years, that "you are what you think." Various accessible individuals in a variety of fields have emphasized the principles of positive thinking and energy; actually they are really emphasizing that you create your own reality. Sir William Osler, the father of American medicine, emphasized that a patient's belief in the physician and the physician's belief in the therapy are more important than what the physician does. Indeed, the single most important factor in accomplishing health probably is, in fact, belief.

The three main determinants of health are: nutrition, physical exercise, and attitude. If everyone ate a natural diet (approximately 75 to 80 percent of calories in the form of complex carbohydrates or starches, 10 percent in the form of protein, and 10 percent in the form of fat) we would abolish 75 to 80 percent of illnesses. If everyone exercised adequately to keep the body limber and achieve a minimum of *Aerobics* points each week, we would improve the health of people in this country markedly. If everyone thought positively and effectively used the mind to think an image properly, we could reduce illness 75 to 80 percent. Even applying 50 percent of the potential in these three major factors, health would be improved beyond the imagination of most people. *Your Right to Fly* provides you with an opportunity to enjoy educating yourself to the responsibility and fun of creating your own reality—and health.

Contents

Dr. James Melton will address your organization upon request. For more information on his other books, and successful cassette tape series contact:

Dr. James Melton
P.O. Box 1991
Palm Springs, California 92263
(619) 323-4204

Introduction

For thousands of years no one flew airplanes. Why? Because no one understood the principles of aerodynamics. Yet today we take flying nearly for granted. Marconi was committed to an insane asylum for suggesting that voices could be transmitted through the air without wires. Today we send not only voices, but pictures as well. Even as recently as the turn of the century, homes were lighted by either candles or gas lights. It took Thomas Edison to develop the electric light bulb and proclaim, "I'm going to make these things so cheap that only the rich will be able to afford candles."

The mind of man is a beautiful thing. It seems that nearly anything that is conceivable is ultimately possible; all that needs to be done is to discover the principle by which it works. Space travel, advanced medicine, nuclear energy, and computers are all recent developments of technology which began as someone's idea — someone's thought. When rightly used, they are beautiful tools which assist man in his physical world.

It is easy to see why a person living in a world of physical convenience would begin to adapt to the technological way of life. It is easy to see how people become complacent and comfortable with the idea of having things done "for" them. Albert Schweitzer was once asked what is wrong with people today, and he replied, "People just don't think!"

Yet man's mind is an amazing tool. If man is left to his own devices, he will discover, develop, and create. But the average person in our society finds that, if he allows it, most of his thinking is done for him. Consequently, many people fall into the trap of either blaming others — "it's their fault" — or defeating themselves—"I can't do it."

I believe that something absolutely phenomenal is locked up within each one of us—something so great and so powerful that we have not seen the likes of it before. As Walt Whitman put it, "You are not all contained between your hat and your boots." I believe that it is possible for each of us to release this energy, either in part or in total, and direct it into any specific area of our lives that we choose.

The problem is that many people do not know how to release this energy. What is even worse, most people don't even know they possess it. Plato said that the truly ignorant don't know they're ignorant. A law cannot be effectively applied until its presence is realized.

My personal experience is that there is a law of life which can be relied upon as surely as the law of aerodynamics. The purpose of this book is to describe how you can gain control of this law and apply it toward your personal health, wealth, and happiness. Although each chapter focuses on a specific area of life, such as sales, body image, and prosperity, it must be noted that in no way is each chapter complete within itself. To try to isolate a problem area and treat it alone would be to ignore the underlying attitudes which cause such a problem. I have often found that when a person is having difficulty in one area of life, he usually needs assistance in other areas as well. Separating subjects like health and love, money and body image, sales and intuition would imply that each has nothing to do with the others, and this is not so. Handling a subject as a specific would be similar to white-washing the water pump and expecting to get clean water.

An attitudinal adjustment in one's thinking must be made, thereby defining the problem area in a more complete fashion. The old saying, "Wherever I go, there I am," is so true. We always take our thoughts with us, and it is by our thinking that we draw to us the world in which we live.

I believe that our awareness that this "law of laws" exists is our springboard to a fuller and richer life, and that by our understanding and application of this law, we have the right to fly as far and as high as our dreams can take us.

James E. Melton

Your Right to Fly

Isaac Newton sat one day
And pondered by a tree
And as he watched an apple fall
He thought of gravity—
What keeps us all upon this earth?
What pulls us toward the ground?
Why don't we float or fly away?
An answer must be found!
The energy which holds us here
Defines this time and place
It focuses our bodies
On a present point in space—
But thoughts transcend the world we know
And curiosity
Can soar beyond the farthest star,
Unlimited and free—
Our spirit brings our minds in tune
With energies on high
No gravity can hold us back
We have the right to fly!

—Martha Belknap
Denver, Colorado

written for Jim Melton

PART ONE

body image

1
Establishing Control

All bodies, the firmament, the stars,
the earth and its kingdoms,
are not equal to the lowest mind;
for mind knows all these and itself;
and these bodies nothing.
—Blaise Pascal: *Pensees*, 1670

It's as Old as Dirt

As a potter, I love to work with clay—feel its coolness ooze between my fingers, stretch it, squeeze it, poke at it here and there. I like the earthy smell, and I especially enjoy seeing the pot being formed on the potter's wheel. The lump of shapeless mass is transformed by the potter's hands into a beautiful form, a work of art.

My very first experience with throwing on the wheel came in 1962 in Milwaukee, where I took a class from Abe Cohen, one of the finest artist/potters in the Midwest. Like most novices I had great expectations of twirling the clay on the wheel and developing a masterpiece that very evening—something that would certainly complement any museum of art. Abe's skillful handling made throwing seem easy, and we listened impatiently to his step by step instructions for the process all of us were to follow.

I left that evening after class a little disappointed. My masterpiece had somehow slipped through my fingers. But I had had a taste of the clay, and although I didn't realize it at the time, I was on the threshold of an entirely new career. Yes, I mastered the wheel, and I can consider myself a potter. I have researched the art, lived with it, sold it, developed glazes, built kilns and taught both privately and in college, and that evening back in 1962 led me to develop my own school of the arts where each year potters came from at least 25 to 40 states and other parts of the world.

In order to make the clay work with me, however, I first had to learn how to work with it—to experiment, study its structural qualities, plastic boundaries, and firing limits. In other words, I had to learn the principles by which clay works and then develop skill at using the material before I could possibly become what we call a potter.

The power of the mind is somewhat like clay. Although intangible, it is nevertheless plastic and it can be molded. It has qualities within it that can be explained, learned, and relied upon. So to make it work for us we must know something of the principles by which it operates. The mind is not whimsical; it follows laws as uncompromising as those of clay, mathematics, or aerodynamics.

Are You Here?

There are two things we cannot doubt: our own existence and the world around us. But what do you suppose is in us that is aware of these two things?

Now, I've always enjoyed logic and it seems obvious to me that any inquiry must begin with a certain amount of order. The first step in any investigation is to establish what it is you're looking for. In other words, sight your target and move in that direction. Striking out to research the Mayan ruins, you would certainly not go to New York to unearth your treasures. The Yucatan peninsula would be much more appropriate. Likewise, if you'd like a certain book and you feel the need to acquire it immediately you wouldn't go purchase a ticket to the football game and sit there all day. No, you'd go to the book store or library.

Why is it then that we hear some people say they desire more fun, more good, more success out of life and yet see them continue to focus their eyes and mind on negative situations—situations that are in no way related to what they would really like or want in life?

Are You Still Wetting Your Pants?

Toilet training a child requires a progression through six steps. 1) The child wets his pants and doesn't care; the mother cares and she changes them. 2) The child wets and doesn't like the feel, so he cries and gets them changed. 3) The child watches as his pants become wet and says to himself, "Now where's that coming from?" 4) He grows to realize that he wets his own pants. 5) He learns to recognize when he needs to urinate, and is able to prevent wetting his pants. 6) Finally, the child makes the connection between drinking and urinating. Moral: Unless the child is aware that he is wetting his pants he has no chance of learning to do otherwise, and he becomes a victim of his own actions.

Growing Pains

A lot of us aren't aware that we're wetting our pants. Many people do not realize that they are responsible for their environment; they see their surroundings as totally external and are unaware that it has anything to do with them. Plato pointed out that the ignorant don't know they're ignorant, and there are a lot of people walking around that don't know they don't know.

Awareness must be introduced—awareness that we can influence our lives by our thoughts. Through this another learning process moves in, that of insights into cause and effect. Outer things are but the result of the thoughts you hold in your head. We cannot think one thing and produce another—yet this seems to be the very notion under which most people function. This faulty notion is the base for all human problems—sickness, poverty, human relations, overweight, etc. The one overpowering truth simply stated is this: you are a magnet, and by your thoughts you draw your world to you. You cannot affect a situation directly without first changing your thought.

A Morning Sunrise

One summer morning I was sitting on the couch watching the sun rise. Coming through a beveled glass window, the early light sprinkled the full spectrum of color over the walls, making beautiful patterns from red to violet.

Before the sun rises, the room is dark. With the rising of the sun the room becomes bright and light, and when the sun sets the room returns to darkness. Now the lightness and darkness are both conditions, one being positive because of the presence of the light, the other negative resulting from its absence. Both the positive and negative conditions, light and dark, are results of the same cause—the sun.

Any condition, situation, or circumstance—be it positive or negative, prosperity or poverty, health or illness, success or failure, thin or fat—any condition can never be the primary cause. To change the condition you must first get to the source—thought. Remember that by your thoughts you draw your world to you. Cause and effect is the law of laws. In this orderly universe nothing happens by chance; everything operates by law. Talking to a group of scientists, Dr. Wernher Von Braun once said that years of probing the spectacular mysteries of the universe had led him to a firm belief in the existence of a higher power—that the grandeurs of this cosmos served only to confirm his belief in the certainty of a creator. Von Braun pointed out that the natural laws of this universe are so precise that we don't have any difficulty building a space ship or sending a man to the moon, and we can time the landing with the precision of a fraction of a second. These laws must have been set by someone. He went on to say, "Science and religion are not antagonistic; on the contrary, they are sisters. While one studies creation, the other studies the creator. In science man tries to harness the forces of nature around him and in religion he tries to harness the forces of nature within him."

There is an old saying, "Know the truth and the truth will set you free." There's only one thing we can be set free from: ignorance. There's only one thing that will overcome ignorance: awareness. Therefore, through awareness we discover the principle by which the law of cause and effect works and then we may choose to use it to better our lives and ourselves.

You're in Charge

You've no doubt heard much talk about the conscious and subconscious mind. It must be remembered that the individual does not have two minds but in fact one mind functioning at two levels. Contrary to the belief of many, we do not necessarily fall prey to all of the whimsical patterns of thought planted in the subconscious mind. In other words, you do not become a slave to the fleeting thought. It is only through the continual bombardment by the conscious mind that the subconscious becomes what might be termed the "taskmaker." It obeys the repeated commands given it by your conscious mind. Therefore, by your habitual conscious thinking you make your own mental laws. By repeating a certain train of thought consciously, you establish definite opinions or beliefs in the subconscious, which make the tasks which are acted upon by the body.

We are in charge of our lives by virtue of the very thoughts we choose to think. It goes like this:

Think the thought.
See the image.
Develop a feeling.
Respond with the body.
Produce the results.

environmental result, circumstance, or situation

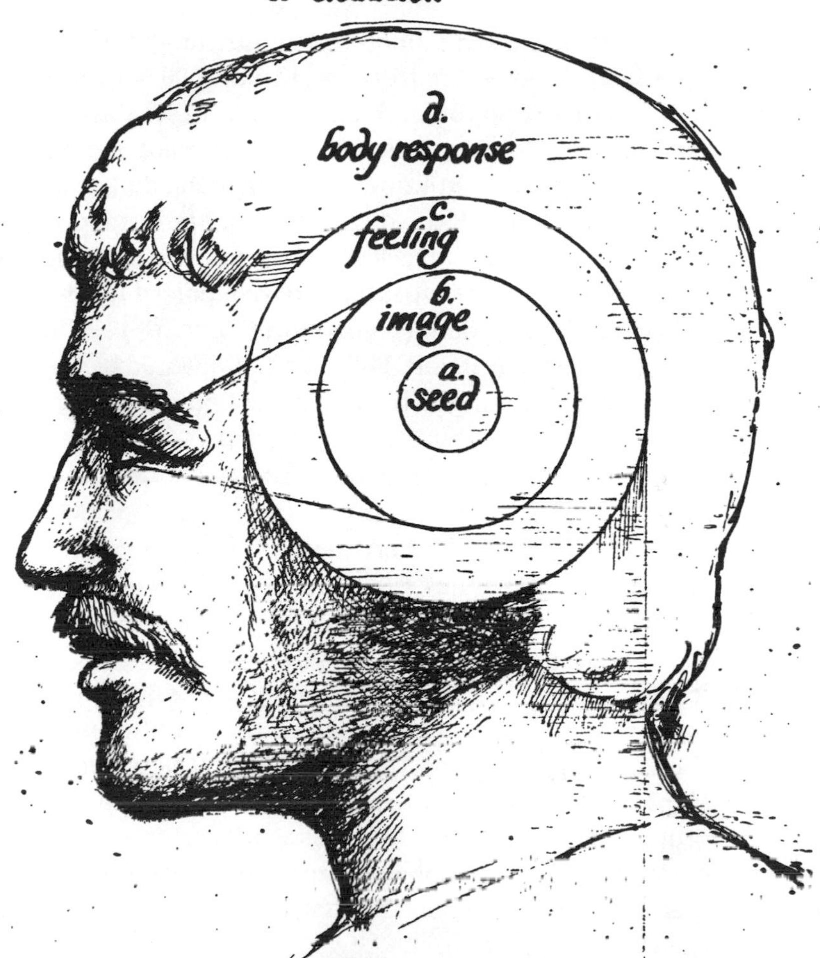

a. Thought seed chosen consciously.
b. image created by your thought.
c. feeling developed in the subconscious mind.
d. body responds to feeling.
e. enviromental condition is the result of the original thought seed.

Let me explain. Choose a thought, any thought, such as a dog. As you're thinking of a dog, I'll bet you a donut to a RyKrisp you didn't spell it out in your mind. You just developed a visual impression of a dog. Chances are you even know what color the dog is, or how it's positioned. If the dog you thought of was your own dog, you probably feel love. But suppose you chose an attack dog—a dog with teeth bared, poised for the leap and then suddenly springing toward you. A feeling of fear may result. Following that would be the final stages of thought in which the body responds, the result being perspiration or uneasiness. Thought, image, feelings, response, results. Try it with anything. Imagine sitting in a large audience waiting to speak to the people. You don't even have to do it to get the feeling. Possibly someone says something and you get embarrassed (red faced) — it happens instantly.

Whatever the mind dwells upon — anything feared or revered—it multiplies, magnifies, and causes to grow, until finally the mind becomes qualified with the new state of consciousness. Fixing your mind firmly on your desires and keeping it there can be quite a job. Can you do it for a week? Perhaps not. A day? Again, perhaps not. Even an hour can be difficult. It seems all we can really do is change our thinking in the present moment.

If, then, we ever hope to better ourselves and our life situation, it would seem evident that we realize that there is no situation or cause external to ourselves that can affect our world. Be it job, family, or money, the thinking process is involved, and as long as we place our success or our failures on something "out there," like that raise, that person, or something like natural talent, then we will continue to seemingly be pushed and pulled, like a puppet, by the world around us. Effective thinking takes practice.

It's Like Playing the Organ

When my good friend Archie Ulm came to Denver to put on a special organ performance, I appreciated once again what a fantastic musician he is—truly a wizard at the keyboard. He can pull sounds out of that organ like you've never heard before. There were about a thousand people in the grand ballroom of the Regency Hotel that day, and Archie performed with such artistry and mastery of his instrument it brought chills to my spine. After the program many people approached me to comment on the quality of Archie's music. I remember one person in particular who came up to me and said, "Jim, that man is terrific. I wish I had talent like that; he's truly a natural born musician." I said to him, "If you had spent as much time and effort, and had as much love and enthusiasm as that man has, 3 to 5 hours a day for over 20 years, you too would be able to do that. No," I said, "he is not a natural born musician."

The Thing Actually Flies

Like anything else, it takes practice to do something well. If you are new to the game of directing your thoughts, if you have formed some "bad" thinking habits, fear not. They can be changed. All habits are pieces of behavior built up over a period of time, and behavior can be changed moment by moment with directed conscious thought. But, like playing the organ, it takes practice and certainly a desire to do it.

Holding on to a new idea requires a certain kind of courage. It is said that when the Wright brothers encountered failure after failure one would turn to the other and say, "It's all right, brother. I can see myself riding in that machine, and it travels easily and steadily." And so it did — eventually. The trick is to keep your mind fixed on a high idea or image rather than on a lower concern. To relearn and repattern your thinking can be quite a task indeed and it does take continued desire and the expectation of seeing results. However, the universe always says yes. It will support you wherever you are and it gives to all impartially.

Oh, That Poor Man

James Allen, the author, once said that men are interested in improving their circumstance but unwilling to improve themselves, and that they therefore remain bound. Frequently we hear people saying things like, "He's such a nice man; why has something as disastrous as bankruptcy fallen upon him?" Or, "How could God let my house burn down? What did I do to deserve this?" When a baby puts his fingers into the wall socket and gets shocked or killed, do you ask how electricity could do that to him? When a boat sinks do you ask, "How could the water do that to that nice boat?" Of course not. Electricity can heat your home, cool your freezer, or kill you; it is impartial. The law of flotation will float anything which is bulk for bulk lighter than the mass of liquid displaced by it; and when you understand this principle you can use its tremendous power. The boat sinks or floats depending on the law, not on whether the boat is nice or not. The bum in the Bowery, the Madison Avenue business executive, the California mother, fashion model, Tibetan Lama all can be supported on water if the law of flotation is followed. So will the law of the universe support each and every one of them in what they do and wherever they go, because it too is impartial. The problem arises when people will not take the time to study the law.

Understanding, then, seems to be a key concept in the effective use of this principle. Often we hear things but we don't hear them. We read a book or hear a tape, listen to a lecture or sit in a class and we learn nothing. It seems that we only hear when our minds are receptive and the presentation is on our level of understanding.

I read the book *Think and Grow Rich* about fifteen years ago, at the request of a good friend of mine. He said I needed it. I did not understand what the book was attempting to say, and I threw it away. Then, seven years later, I read it again, and this time my response was, "Where has this been all my life!" You see, I was ready to hear. The book was the same, but I had changed.

With this in mind, I'd like to attempt to explain an important factor in understanding this life principle. More than that, I'd like you to be able to use this in your everyday life. The following three examples show why I believe you already have everything you need to be successful.

Wash the Windows

It seems to me that we don't have to *add* anything to our lives; instead, we need to *take away*—take away the junk. The other day I was washing the windshield of my car. It always pleases me to see the fresh, bright, clear windows so I do it often; besides, clean windows make it much easier to drive. I'm sure you've experienced the frustrations I have while driving down the road with a dirty windshield. The sun is beaming and you can't see a thing ahead of you. The particles of dirt distort and disperse the sun's rays in various directions, causing you a real problem in driving. My own simple solution is to take the dirt off the windows by washing them often. You see, the glass is already clear, but the dirt has to be taken away.

The same is true with water. In order to make it drinkable the impurities must be removed. Nothing need be added; the water just needs to be cleaned up a bit. Nature has one of the most effective recycling systems ever developed. There is no new water; the water is all recycled, day after day, year after year, century after century. You may be drinking the same water as Jesus, Napoleon, or Cleopatra. The water itself is clean; it just needs the dirt removed.

Have you ever seen a rough-cut diamond? It is not really too beautiful; it looks much like any other stone. Then the diamond cutter cuts the facets on the gem and all the brilliance, radiance, and beauty are released — through the facets. Each facet expresses the total vibrance and beauty of the diamond, not one more than another. And the diamond cannot withhold any of this from the facets. However, when the facet gets dusty or dirty the residue must be taken away for the radiance of the diamond to be released.

For all of these—the windshield, the water, and the diamond — nothing needed to be added; rather, something had to be taken away in order to bring it back to a perfectly functioning state. You might consider that we are the facets of life, and life can no more withhold energy, health, good, or love from us than the diamond can withhold any of its qualities from the facets, *unless* we cloud ourselves with fear, self-defeating behavior, or doubt. Only when these negative qualities are absent can the true nature of our being be exposed to the world.

What you concentrate on will tend to manifest itself. The principle of aerodynamics was not discovered by contemplating things such as dogs, trees, and rocks which don't fly, but by contemplating birds which fly naturally—their structure, shape, and weight — and then posing the question of how they were able to do so. Insight into any principle is gained by studying its positive action. It would seem, then, that to live more successfully a person must study the successes in life instead of continually observing the failures, fears, and disappointments. To do this would certainly promote advancement in anyone's life.

Believe Me!

You say you can't change because that's just "the way things are?" One of the greatest laws of mind is that of belief. You are the sum total of the thoughts you choose to think, and the thoughts you choose to think form your beliefs. Now, did you get that? I didn't say that you were your beliefs; that's not true. It *is* true that we function in accordance with our beliefs. There is a difference.

We hear so many people say things like, "I'm fat," "I'm a lousy cook," "I'm no good with mechanics." The fact is that they believe that about themselves and therefore they experience that exact thing. Take, for instance, the person who believes he is no good with mechanics; auto mechanics to him is a complete mystery. If he sits down with an expert auto mechanic, someone who loves the field; if he reads some good books on the subject and goes to school to learn auto mechanics, lo and behold he's developed himself into an auto mechanic. How? Through choice and training. He has changed his belief from "I'm no good with mechanics" to "I'm a good auto mechanic."

You see, you are not your belief; rather, you are the person who experiences those beliefs. The answer in redirecting or completely changing a belief lies in these two things: choice and training. You must first desire to change. If you don't really desire to change, why bother to try? That makes sense, doesn't it? So, if you do desire a change in belief the next step is to train yourself both mentally and physically in accordance with your new belief. "I'm broke," you say, "and I believe I'm broke because I experience it!" You are not broke because you don't have any money; you don't have any money because you believe you are broke.

Looking for Help?

When a friend of mine quit her job at a bank to take what she considered to be a better one, she went from a secure position with a generous salary to a job that gave her more freedom of expression. As a sales job it's strictly on a commission basis, but for her it offers more opportunity for growth. I admire her courage in moving from security to insecurity; she believes in herself and knows that if she is not happy where she is, change will never occur unless she takes action. In other words, we can't expect our lives to change by continuing to do what we've always done.

Suppose you were the bank's personnel manager, looking for someone to fill my friend's shoes. What kind of person would you like to hire? If you could sit down and make a list of all the qualities you desire in an employee, what would you list? Wouldn't you want a creative person, someone dependable, conscientious in attending to all matters, able to follow instructions, and without a doubt totally honest? Someone willing to work long hours, efficient, bright, capable, and always agreeable? Someone who enjoys the job so much that pay isn't important?

If you were fortunate enough to find this ideal employee, you would not ever consider ridiculing him in any way, would you? No putting him down, no talking about him negatively behind his back. You might even be concerned about the people you introduce him to, making sure that they are top quality people in every sense of the word—because you know that people tend to join into whatever type of conversation is at hand and you wouldn't want your employee to be negatively influenced. You'd be caring and understanding, and you would probably take whatever time was necessary to explain any changes in procedure along the way, knowing that this would be a one-time explanation because this guy also has, among all his other qualities, a perfect memory.

Too much to ask in an employee? You are wrong! It's not too much to ask. This person exists right this very moment within you—in a portion of the mind we call the subconscious. Each person is walking around with one of the most powerful, faithful, creative, and obedient pieces of equipment in the universe: the mind. But if we don't recognize this, we can't use it. It's like leaving a prospective employee standing in the bank lobby waiting to be interviewed for the job. If he doesn't get interviewed he never gets the opportunity to take on the challenge, and therefore he can never perform for you. If you could give your subconscious mind a chance to be interviewed, you would find a wealth of assistance to handle the everyday problems that come up. But the big problem is most people won't take the time for the interview even when they believe it's possible.

The Amazing Taskmaker

The perfect memory, the instant doer, the intuitive calculator, the most highly sophisticated computer on the planet, the amazing taskmaker — your subconscious mind.

So you think you have a poor memory? You may be in for a surprise.

The first guest on Johnny Carson's *Tonight* show one night was Harry Lorayne, author of *Secrets of Mind Power.* Lorayne is a memory expert. Just before the show, he had asked each person in the audience to tell him their name. Then, when the show began, he had all the people in the audience stand up and he proceeded to say their names, one by one, first and last, never missing a single one. He named 750 people, each by first and last name. Could you do that? The answer is—most definitely yes! You probably have not trained your mind as effectively as Harry Lorayne, but the capability is there if you desire to use it. The difference between the two minds, yours and Harry's, is not that one has a greater capacity than the other. The difference is that Harry's mind has been disciplined, trained to perform such feats. While he uses such things as memory pegs and imaginary patterns and the like to assist him, the fact is that he has a fantastic memory because he has trained his mind in that area. He may be unskilled in other areas, but unquestionably he is a memory wizard and he believes it.

What did you have for breakfast ten days ago? If you're on a diet you may remember, but the chances are that most people haven't the slightest idea what they ate for breakfast yesterday, let alone ten days ago. Certainly, it's not important. However, it is interesting that in a state of hypnosis a person, when asked what he had for breakfast on this day ten years ago, will without hesitation tell the hypnotist exactly what he ate. He'll also be able to remember what he was wearing, what the weather was, what his feelings were, and who he was eating with. We don't forget things; we just tuck them away for later use or disuse. It is strictly our choice.

Another example of effective memory is a study that was done regarding reading. A man was asked to read a paragraph in the middle of a newspaper page. Upon completion he was asked to recite the complete paragraph as best he could. He was able to do quite well in getting the general theme but could not recall the paragraph word for word. Then, under hypnosis, he was asked to repeat the paragraph, and he did so without error. *Then* he was asked to recite the entire page, which he did not read, and he did so!

Why was he able to do that? He didn't even read the page of print; how could he know what it said? The answer is quite simple. We take into our minds much more than just what we think we see. Our peripheral vision reaches the brain just as our direct vision does, and we remember what we saw with it, too. Carry this phenomenon a step farther to hearing and touching. Do you suppose you are actually hearing the conversation over in the corner of the room, even though you are actively engaged in a conversation with someone else? Whether distant voices and visions are perceived by our outer physical senses or our inner intuitive senses, these impressions do seem to make their way to our subconscious and remain there, stored for possible future use.

Another example of peripheral visual bombardment was obtained through an experiment in which a woman was asked to walk down the city street. Upon returning she was asked what she saw in the department store windows. She was able to mention a few of the items but her description was quite general, her comment being that she didn't know she was supposed to make observations in the windows. After being placed in the hypnotic state she was able to describe in detail the items in the windows and even specify which windows they were in.

Your mind is perfect. You have a perfect memory and if you choose you may develop it to its fullest extent. The mind is bombarded with hundreds of thousands of impressions daily. Many of these impressions are so slight that they slip right past the conscious mind and implant themselves directly into the subconscious. It is this foundation on which subliminal advertising is based.

A study investigating the effectiveness of subliminal advertising measured the amount of popcorn purchased by the viewing audience during a normal run of a film in a local theater. Then, into the same film were spliced, at random spaces, single frames of the word popcorn and photos of popcorn. A sound film runs through the projector at 24 frames per second. In normal operation the rapid aperture closing of the projector produces an image only every

1/48 of a second—not enough time for the human eye to catch it or recognize it for what it was. However, during the running period of the spliced film, the audience viewing the film consumed seven times more popcorn than during the previous film. Good for business, wouldn't you say?

Subliminal advertising leads to the question of whether we buy products because we like them or because we have been programmed to like them. Sometimes advertising is not so subtle. For instance, you may find an item in your grocery bag not because you really can't live without it but because John Movie Star uses it. The key here is awareness—awareness of making your own choices, of not acting because you were told to, forced to, or misled. Maybe you want to help support John Movie Star; maybe you really like the product or would honestly like to try it. But to buy without thought is to give up control over a part of your life.

Natural Hypnosis

Many people are afraid of hypnosis when they are, in fact, hypnotizing themselves every day. It's a state I call natural hypnosis. Natural hynposis comes about through habit—continually reacting to situations in the same way, continually viewing the same people in the way you always have, carrying yesterday's thoughts into today. Affirming over and over again that "that's the way things are; they never change" produces natural hypnosis.

A continual bombardment of ideas into the subconscious will develop a believed pattern of thinking. When the subconscious mind believes something, it acts on that belief, even if it's something you don't happen to like. That's why I believe so strongly in a process I call directed conscious thought. By consciously directing the thoughts that you think, you can determine the actual circumstances you will encounter. Remember, your world is magnetized to you by your thoughts. Van Dyke says that "we create our fortune thought by thought." James Allen, author of *As a Man Thinketh*, said that:

> *Mind is the master power that molds/*
> *and makes—man is mind.*
>
> *And evermore he takes the tool of thought/*
> *and shaping what he wills*
>
> *He'll bring forth a thousand joys or/*
> *a thousand ills.*
>
> *We think in secret but it comes to pass*
>
> *Environment is but our looking glass.*

Learning to direct your thought is like learning anything else; it takes practice. A process used in the Far East consists of observing your breath as it passes your nose on inhaling, then observing again upon exhaling that same point at which the air passes your nose. The idea is to focus in on one point and be conscious of only that point. This method is taught by yogis, who recommend that the student practice many hours each day for months.

Another method of directing the mind is with the use of a mantra—a word repeated over and over to the point of boredom, thereby producing an altered state of consciousness. Still another method of directing your thought, and more enjoyable than the previous ones mentioned, would be relaxing in a comfortable place in your home, closing your eyes to rid yourself of visual stimulation, and then gently observing the stream of consciousness that goes on in your head. When you notice something drift in that is not conducive to the state of being that you would like to express, gently observe it as it flies across the hills and valleys of your mind and right out the other side. Don't force it out; just watch it float through, in one side and out the other, like watching a movie. While doing this, every once in a while deliberately choose a positive statement or image that fits in with your newly desired lifestyle. Plant it and watch the follow-up patterns in the mind. Play with it; make a game out of it. Imagining that longed-for trip to Hawaii could be the initial thought. The game is to carry it further by involving all five senses. Smell the warm humid air, the fragrance of the flowers, feel the coolness of the water flow gently over your feet as you mentally stand on the sandy beach. Hear the sound of the waves beating against the shore. Form a mental picture of you actually being there. Involve your sense of taste also by tasting the salt air or possibly imagining the sweet juices of a pineapple but above all, make it enjoyable.

This is directed conscious thought in the freest, most expressive form I know of, and leads to the truest form of meditation or knowing type of thinking that I know of. Caution must be taken not to fall off to sleep; this will take practice also, for it seems that the only time people really relax nowadays is just prior to sleep at night.

Life directed thinking. This is a pattern of thinking which becomes so habitual that as you move through your day, working, playing, resting, your thoughts are screened through a strainer of life-directed attitudes, survival attitudes. You consistently take in thoughts, both positive and negative, and sift them through your life-directed strainer, sending them back out into the world with a new form of positive energy. You transmute everything that comes to you to fit the pattern or lifestyle you would like to see manifest itself in your life.

Physical vs. Feeling (Emotion)

Directing the mind is certainly nothing new. We've all been doing it all of our lives. Learning to walk as a child, controlling the food in our mouths, driving a car, playing tennis—all are physical skills and in order to do them properly the mind must be directed into that area. Why is it so complicated, then, to suddenly begin to take charge of your thoughts in the emotional feeling area? (There are tasks at hand that need to be undertaken there as well as in the physical area.) We learn the physical tasks to the point they become habitual and give them constant reinforcement all our lives, and do not consider that the same techniques can be learned for mental and emotional tasks as well. Is the possibility that we can actually be in control of all aspects of our lives a totally new discovery? Actually, it's not! This kind of control has been around for years; it just hasn't been applied in the area of the mind as much as the physical realm. Just as we learn not to put a hand on the hot stove or to avoid a protruding board on the back porch, so can we learn not to be unhappy, not to be angry, and not to feel guilty.

Yes, taking charge of our thoughts is the law of laws. It is cause and effect at its prime. Remove that foggy veil of ignorance and a whole new world opens up. In a case of happiness vs. unhappiness, who would willingly choose unhappiness? Surprisingly, many people do. Knowing the way is not necessarily going the way. You may know all the right things to do to better your life but be absolutely unwilling to give up the habits — the trash — that go along with failing. Believe me, that's actually a way of life with some people; such thinking has become a hard and fast habit. But habits can be changed. Although it's easier to remain in the old patterns than to change to the new, if the new are better the rewards are worth the effort. Just as habits were formed, they can be broken.

Having read this far you are now somewhat enlightened on the workings of the mind. For many people, saying "I won't use this stuff" is like refusing a glass of water after a ten-mile walk in the desert.

Right here I'm going to insert a word of caution for using the power of mind, and I suggest that you read this part carefully. It is possible to end up with egg on your face with this newly developed power.

A friend of mine, Sue, was playing tennis with her husband and another couple. They had just finished one match and their next match was to be played with a couple considerably less skilled than they. Sue boasted to her friends, "Now I'll really show you how this positive thinking stuff works. Just watch me on the court." You guessed it. Sue lost. She was crushed, royally. She humiliated herself in front of her friends and became extremely uncomfortable around them. She also did something even worse. She began doubting her abilities regarding the mind. Could she in fact use her mind to benefit her life, or was it all a big farce? At that point she actually didn't know.

Another friend of mine, Mary Lee, was losing weight through using the power of mind. At first she lost a goodly amount—and then began the speeches. "It's easy; all you have to do is this, this and this!" "Don't bother with your diet, do what I'm doing!" Well, she put it all back on plus ten more pounds to boot, and she was down in the dumps for months, feeling defeated, humiliated, and unloved.

Some of my own experiences have been similar. When I first took up this way of thinking I turned more people off to it than on to it. Why? I was too cocky. I'd found something new in my life, and although I didn't fully understand it I became a teacher, going directly from elementary school to Ph. D. It didn't take me very long before I learned a very important rule, and that was, *tell the world, but show it first.*

It's obvious to me by the results in my life that I have come a long way in working with the law, but it is also obvious to me by the results of some other things in my life that I don't understand everything.

I can't stress strongly enough the need to go your way quietly, doing what you feel is right, challenging these ideas by experiment. As far as I can see there is only one way to really pass on these ideas to someone else, only one way to impress upon your family, friends and loved ones the importance of proper thinking. That "one way" is by modeling. That's right; in order to stress that successful living is possible for everyone you must live it. I believe that if successful living isn't visible from the outside, you don't have it inside.

Remember, too, that you can't tell anyone anything unless that person is ready to hear. You may have heard the saying, "The teacher will appear when the student is ready." We are all teachers and we are all students. We learn from each other and it is up to each of us to apply what we've learned. No one else can do it for us.

Fifteen Years at It

Driving a friend home one day, I listened to her telling me of the new book she was reading and of a new class she was about to begin the first of next week. Such readings and classes were nothing new to her, and she was full of advice to others for solutions to their problems—even though her life was an absolute mess. I asked her how long she had been studying these ideas and she said, "Oh, Jim, I've been working at this for some fifteen years now." Right then it hit me. It's not how much you know; it's how much you *apply* what you know.

In my studies I've found out that there are a lot of courses of instruction and a lot of books and tapes that will take us on great esoteric trips. But these don't put bread in our mouths, make our bodies healthy, or put money in the bank. And unless these ideas do those things then they aren't really valuable to us.

Reading books, going to classes, listening to tapes all mean nothing unless we do something with the insights they provide. What good is it to gather together a room full of money if you're not going to spend it? And what good is it to gather together a head full of knowledge if you'll never use it, or worse, if you are going to condemn yourself for not using it? The results in your life tell exactly what's going on inside. When you start to see results outwardly for actual and practical use only then can you tell whether you have actually incorporated the ideas which were studied. It is the understanding and application of wise thoughts that counts.

2
Flyin' High

We are told what fine things would happen
if every one of us would go and do
something for the welfare of somebody else;
but why not contemplate also the
immense gain which would ensue if everybody
would do something for himself?

—W. G. Sumner: *An Examination*
of a Noble Sentiment, 1889

Almost anyone—salesman, lawyer, bookworm, farmer, executive, tailor, mother, hair stylist—almost anyone cares about his or her physical appearance. People often like to think they look like Betty Grable or Farrah Fawcett, Clark Gable or Robert Redford, and the beauty business indicates how widespread is the attempt to buy the way to beauty, with weight loss programs, pills, shots and diets taking the lead, trying to sway the nation into thinking it's natural for all people to look like a model in *Vogue* or *Esquire*. Granted, the models do look good on the page, but most people are not models.

This trend of thinking is illustrated by "if only" thinking: "*If only* I could lose this extra pound and a half, *then* I'd be beautiful." If I've heard that once, I've heard it a hundred times. I have conducted weight loss programs for several years, and I do know the importance of shedding those unwanted, unattractive pounds, but to pin beauty on losing such a small amount indicates how thoroughly we've been brainwashed.

Teenagers, not prey to the pressures of battling the bulge, are bombarded with advertising to clear up their faces—as though it is a crime to have pimples—and the gray-haired are advised to cover up their *aging* gray. Add hair, remove hair, take it from one place and put it someplace else. Spray your hair, spray your mouth, spray your armpits, spray your genitals, spray your feet, powder your body, perfume your ears, cologne your face, fruit-flavor your lips—but above all don't let anyone ever see, taste, smell, or touch the real you.

When little Johnny goes off to school, he is told to comb his hair, brush his teeth, wash his hands, button his shirt, say thank you, and be quiet. In most cases there is no concern for dressing up the thoughts in Johnny's head. It is strange that many people spend hundreds of dollars each year dressing their outer appearance and often spend not a nickel for the inside. Most people interpret this education as meaning to fill a child with information. Actually, the word educate is derived from the Latin word meaning to bring out from within.

Marcus Bach, one of the foremost authorities on world religions, mentioned during one of his talks that while in India he talked with a Buddhist monk and asked why not many of the young people attended the morning meditations. The Buddhist answered,

"We believe that the young people need less teaching and more examples." He mentioned also that "if we make our religion a very important part of our lives the children will see this and choose for themselves the right direction." I heard it put his way: "Someone is following in your footsteps. Are they worth following in?"

Every loving parent wants his children to be free expressions of life, living it to the fullest, but we fail to realize that children learn from the behavior of their models—the parents. We say things like "be your own person, don't follow the crowd, think for yourself, be number one, excell, achieve, feel good about yourself." Remember, they are following not what you say but what you think and do. Are you number one in your book? Do you feel good about yourself? Are you your own person? The way you view yourself is the way the world, including your children, views you.

The late Dr. Maxwell Maltz, internationally famous plastic surgeon and author of self-help books, said that he changed the outward appearance of many a person by removing scars, birthmarks, and the like, but unless on the inside, in the person's own mind, the person decided to feel good about himself, the outer change was of little or no value. Maltz indicated that it was as if personality itself had a face. This nonphysical face of personality seemed to be the real key to personality change. If it remained scarred, distorted, ugly or inferior, the person himself acted out his role in his behavior regardless of the changes in physical appearance.

Losing weight provides a good example. The most common thing to do if you desire to lose weight is to go on a diet. Some people resort to exercise or pills; some take injections. Often when a person is unsuccessful he turns to his family and friends to monitor his food intake: "If you see me take just one bite of this, that or the other thing, I want you to take it out of my mouth." For extreme cases there is the bypass operation; wiring the dieter's mouth shut has also been successful.

All of these methods will work, of course, but only for a short while. The notable thing about every one of these weight loss methods is that each deals with something done on the outside—some physical thing you do. Not one of them deals with the cause of overweight: *your thinking!*

Everything in the physical world has its origin in the nonphysical. You can't get much more nonphysical than thought, now can you? The chair, the table, the light fixture, the clothes you're wearing were all but an idea at one time, and so it is with your present body condition, be it overweight or whatever. Today's body is simply a physical manifestation of yesterday's thoughts. If it's all so simple, then why is it so difficult? Let me explain.

Not long ago I was in Los Angeles to do one of my programs. I was sitting in the hotel lobby and a large, round man briskly bounced into the room and sat down. His first words were, "Whew! Am I tired." Then he looked around for a response from the other people—he didn't get any. So once again, "Man! Am I beat." And again he waited. Now there are three reasons why I believe you should never tell anyone why you feel bad. One is that it definitely reinforces the behavior. Second is that most people don't care. As much as you'd like to think that they do — they don't! No one likes to be around people who complain. And thirdly, some people are actually glad that you feel bad. The reason is that it places them just a step above you so they are now in a position to give advice.

Now, back to the round man. I finally asked him what was wrong. He said, "Today has been really rough. I had to pick someone up at the airport, drive all the way down to San Pedro, back up to Compton on the freeway during the rush hour, and then — on top of it all — my air conditioner went out!" I said, "You don't like driving, do you?" He said, "You're kidding, I hate it!" I asked him what he did for a living and he said, "I'm a cab driver."

Why is it that people put themselves into uncomfortable situations? In many cases I believe it is because they are misdirecting themselves. In other words, they are feeding in the wrong direction.

Quite possibly, if you are not where you'd like to be — if you are not as happy as you'd like to be, or your body is not the proportions you desire, or you aren't as vibrant, as youthful, or as loving, caring, or thoughtful as you'd like — it's quite possible that someone has been giving you, or you have been giving yourself, the wrong direction. In any event, you've been filling your head with non-survival ideas. These ideas are detrimental to your very existence. They are failure-directed and self-defeating, and tend to keep a situation in a non-working status.

Psychologist David McClelland of Harvard says that we talk to ourselves all the time. That's okay, he says; the problem is that we believe what we say. What are *you* dumping into your mind?

I'm nervous | *Everything I eat turns to fat*
People don't seem to like me | *I just smell food and gain weight*
I'm losing my memory | *I shouldn't have eaten that*
It runs in the family | *I wish I could eat that*
That's too good to be true | *I shouldn't let that go to waste*
That gives me a pain | *This is fattening*
Oh, my aching back | *I ate too much*
I'm tired | *I'm starved*
He makes me sick | *I can't do anything about it*
It's a cut-throat world | *I guess I'm meant to be that way*
I can't afford that | *My body is so efficient it wastes nothing*
Easy come, easy go | *I just look at food and gain weight*
He gives me a pain in the neck | *There's only so much I can do*
She drives me crazy | *I can't stand the pressure*

Sound familiar? Words, you say, only words. Yes, but words are an expression of what you think, and we think in pictures and feelings. We always think in pictures and feelings, and what's more we respond to what we think. Imagine scratching your fingernails firmly on a blackboard. How does that feel? The same thing applies to reliving an embarrassing experience —when you get good at it you can make it even worse than the actual experience.

Dr. Edmund Jacobson proved that the body responds immediately upon thinking something. In one experiment he had a telegraph operator tap out the Morse code while hooked up to a sensing device with an oscilloscope readout; Jacobson was able to observe the record of the operator's movements on the screen. Then he asked the operator just to think of tapping out the Morse code. Merely the thought of it activated the little muscles in the operator's arm enough to give a readout on the screen. The body responds instantly to your thought.

When we say something like "I'm tired," "I'm exhausted," "I'm worn out," the body responds beautifully to our command. We plunk down in a chair and can't seem to get up again. What could we replace those statements with? What could we say without lying to ourselves and pretending that we're not tired? How about "I'd really like to refresh myself!" Or, "I'd like to relax; a break would sure feel good." Any of these can be substituted to direct the flow of energy for your benefit. What we need is a diet all right! A diet from negative thinking.

Still not convinced? The fact that you're with me this long gives us both a clue that you are really interested in developing your awareness. I heard a young man who had just graduated from the university with his Ph.D. say, "I have completed my education." I thought to myself, "Son, you haven't even begun." During a talk given by Ann Francis I heard her relate a story about her little girl. She was saying her prayers one night, and on completion said, "I love mommy, I love daddy, I love my kitty, but most of all, I love me." She turned to her mother and said, "Do you know why I love me, mommy? Because I'm here, I'm going to school and I'm learning." Isn't that what it's all about? We are all here, going to school and learning. When we consider that school's out and we're learning no more, there's no need for us to be here. Richard Bach puts it this way: "Here is a test to find whether your mission on earth is finished: If you're alive, it isn't."

Start thinking of your body in a different way. It's a vehicle, a sensory device, a three-dimensional stereophonic TV set with a digital computer for a brain. In fact it's the only possible way for you to experience life on this plane of existence. Got that? The *only* way. Everything you see, smell, taste, feel, hear, think — anything that you can possibly do or think of doing must go directly through your body. It takes you places, meets people with you, smells a beautiful rose for you, views those spectacular sunrises and sunsets. It does all those things and even more.

But we must remember that although our bodies are great, we are not our bodies. Many people think they are their bodies, and that's the basis for many of their problems. Your body is not you.

Mirror Mirror on the Wall...

I got up one morning, stood in the bathroom in my all-together, and looked into a full-length mirror. I said to myself ... "Who are you? Why, I'm Jim Melton of course." But my mind snapped back and said...

Mind: No you're not. Back on the West Coast many people called you Mel, and people have called you many other things. No, you are more than a name.

Me: Then, ahhh, I must be my body.

Mind: No, wrong again. You don't phone into work and say that "body won't be in today, boss," do you? Remember Bill Barnette? He returned from Vietnam minus a leg. Did all of Bill return?

Me: No, he was missing a leg.

Mind: Yes, but was he any less Bill?

Me: Of course not; he was still the same old fun-loving guy as when he left.

Mind: And Tom Cowden, the cabinet maker. Do you remember when he began his career, young and careless, and he sawed off a few of his fingers? Was he any less Tom?

Me: I see what you mean. But if I'm not my name and I'm not my body, then who am I? My emotions? No, don't answer that. That can't be it either, because the same me experiences both happiness and sadness, excitement or depression.

Mind: How about your job? You hear people saying "I'm a plumber," or "I'm a doctor."

Me: No, I couldn't be that either or I'd be different people all the time. As often as I change jobs, I know the real me couldn't change that often. All right. My name, body, emotions, job, I guess even my marital status are not me, because all that can change, and the real me experiences all that stuff.

Mind: Yes...experiences it all through your body.

Standing in front of the mirror, I found out what I was not, and when I found out what I wasn't, I realized what I was. I am mind, expressing myself as Jim Melton.

The Beach Boys

When I lived in California I went to the beach often, and on occasion I would choose Muscle Beach in Santa Monica. I enjoyed watching all the body builders, both male and female. The push ups, chin ups, flexing, pressing and all that goes with body building was a sight to behold. These people view their own bodies all the time; they are body worshippers.

How do you suppose these people look at themselves? As undernourished, peaked and unattractive? Of course not. They see themselves as vibrant, energetic, and healthy—body beautiful. The picture they hold in their minds is that of a beautiful healthy body. What kind of picture do you hold in your mind of you? *The way you greet yourself is the way the world greets you!*

Are You What You Eat?

Getting the body in shape is one of the first thoughts on people's minds these days. I'm not going to tell you what you should eat as a daily diet; I can't, as you'll see shortly. I can only make you aware of the fact that certain types of thoughts and emotions sponsor a desire for certain types of foods and drink. You see, I personally don't think that anyone can tell you what your exact balanced diet should be unless he has some kind of readout as to the thoughts you hold in your own head, or at least a general trend of thinking. I say this with confidence because I have found general patterns woven through thousands of people in my body image classes and other programs I've investigated.

A Three-Legged Stool

I begin on the premise that we are all body, mind, and spirit. In a sense we are always in tune, because we can do something physical and it will affect our mental and spiritual aspects. Likewise, with your thoughts you can affect your physical body and your spiritual side. In other words, changing one automatically affects the other two. It's like a three-legged stool. You can't cut a piece off and have a balanced stool; the other two must be chopped down to size also.

The press of life seems to be for expansion and fuller expression, and as all things in life we see balance, a blending of two to bind together as one unit — up and down, in and out, black and white, day and night, action and reaction, hot and cold. The universe and all it contains is in a constant state of flux. You might say that we live in an ocean of motion, where everything moves, everything vibrates, and nothing rests. All things are continuously changing, and in fact that seems to be the only constant thing: change.

Yin and Yang

The Orientals call this universal balance yin and yang. Some synonyms for yin would be passive, negative (as in negative pole of electricity), resting, saving up, inflow, often considered a feminine quality. In foods we notice that yin foods contain more potassium, while yang foods contain more sodium. On the yang side we have the active, positive, spending, outflow, often referred to as masculine. Any power, as you can see, is not single but dual by nature. Everything in the universe is a matter of balance, and it is constantly moving and balancing. When two poles of a power are complementary a balance is present. A state of perfection exists — a great strength of the power functioning. Marriage, for instance, is only effective when there is a complementary situation, a true balance present. Often in marriage the male will be expressing too much yin, or the female too much yang, thereby causing an uncomfortable situation. Each may then choose to seek their complement in another partner.

Often there seems to be no choice, and we believe we are victims of circumstance. However, through the yin and yang concept we are offered yet another choice. With understanding and awareness it is possible to see deeper into life, enabling us to remove some of the pain from painful experiences, be it overweight, marriage conflict, or sagging self-esteem.

Everything and everyone contains both yin and yang, at all times. They always interact; there is not a single moment of rest in their rise and fall. They are neither good nor bad. They just are. They are equilibrium and harmony as well as conflict and opposition. And again, the best way I can describe it is the swinging pendulum. Picture that or possibly the seasons. Spring and summer are yang, getting ready to do and then doing. Fall and winter are yin, getting ready to rest and then resting.

The following is stated so beautifully that I wanted to include it; however the source is unknown.

> *The Great Ultimate* [*principle*], *through movement generates yang. When its activity reaches its limit, it becomes tranquil. Through tranquility, the Great Ultimate generates yin. When tranquility reaches its limit, activity begins again. Movement and tranquility alternate and become the root of each other.*

According to this passage, one extreme causes the other. Perhaps it's a need for the opposite to balance where we are at the moment. There is a mental equivalent for everything—a related vibration. So it is with yin and yang. Thus, if you are involved in a very yang environment — lots of things that spend your energy, such as anger, noise, confusion, worry, indecision, too much to do, frustration — you will probably crave the balancing yin — total solitude, quiet, rest, meditation, sleep, relaxation.

A World of Sweetness

Look at the chart of yin to yang foods. If you've had an environment of nothing but yang attitudes and activities, what do you suppose would be a physically balancing thing to do? You guessed it: eat sweets. You eat the sweets, causing you to lower your vibration (sugar being a natural depressant) toward the yin. This is natural for all people. However, some people carry this one step further; overweight people. They eat the sweets, which are very likely on their "no, no" list and promptly condemn themselves (mental yang), swinging the pendulum right back to yang thereby causing a desire again for yin. Now isn't that sweet.

Since we are all body, mind, and spirit, a balancing activity could have been entered into instead of eating sweets. (Please realize that eating the sweets did, in fact, accomplish the same goal of slowing you down so you would *desire* to swing away from the extreme yang activities in your life.) What your body is saying is "I need a breather."

Look at the yang balancers for boredom. All that stored up energy needs to be spent and those foods

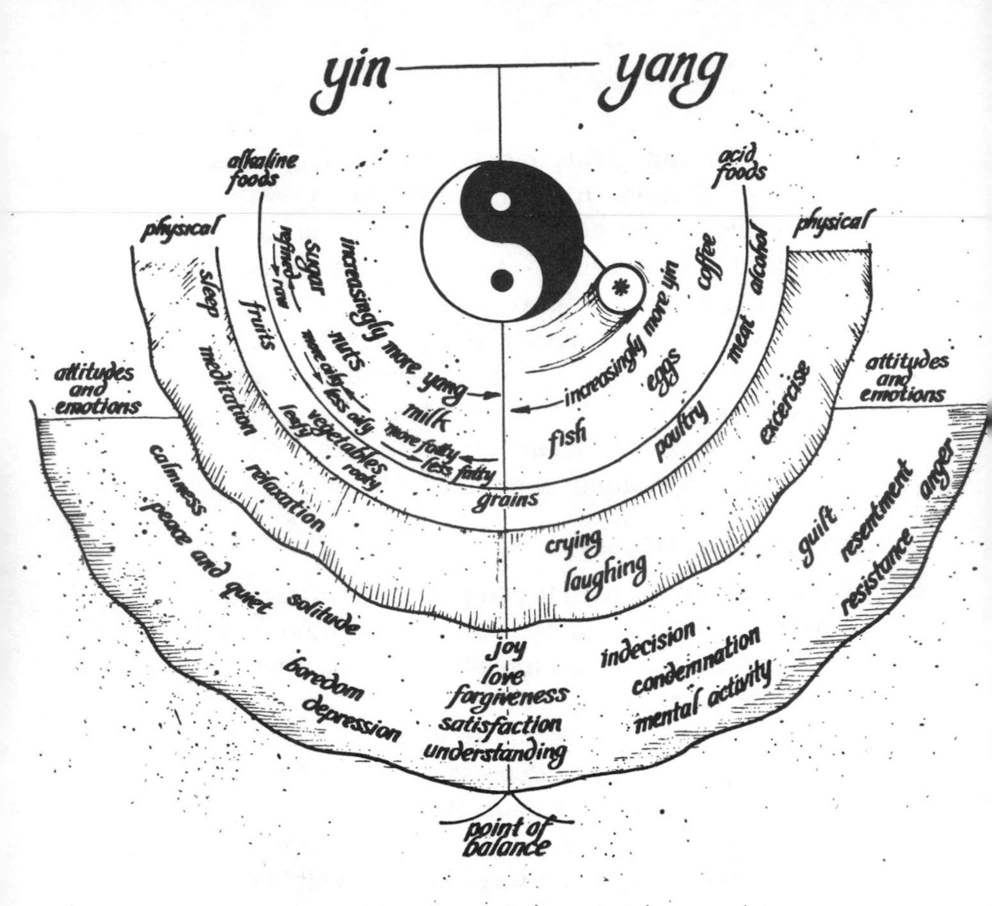

The above figure provides a general classification for yin yang foods, activities, and attitudes. A more specific breakdown may be determined by considering environmental conditions and chemical composition of foods. Cooking may greatly change the food qualities from yin to yang and vice-versa.

* *Life is in a continual state of motion (as indicated by the pendulum). Through awareness of where we are, we adhere to the philosophy of "All things in moderation." Thus a comfortable balance is achieved. A swing of the pendulum to one side automatically demands a swing to the other side — life does not stop!*

will certainly make you feel like spending! We too often use spending emotions like self-condemnation instead of an activity that adds to our self-esteem—an activity which is constructive or self-fulfilling.

I used to be one to cause a great deal of yang in my world by being extremely hard on myself about organizing and getting things done. Not accomplishing enough, in my opinion, was a capital offense. My mind would go non-stop for hours on end (mental activity—yang). Finally, I'd listen to my body and I'd take a nap. But here's the clincher. When I'd wake up I'd slam the pendulum off the map by being disgusted that I slept when I should have been doing something. Then depression would come in and I'd be bouncing back and forth between two extremes. I didn't understand yin and yang at the time.

Four Dozen Cookies

Gwen was going on a three-day fast, to lose a few pounds of course. Now I recommend that you listen to what your body desires and eat accordingly. It was suggested to Gwen that if she was going to fast it would probably be wise to at least keep some vegetable and fruit juices around so if her body really did desire them they would be available.

She started off on Monday at her job, a mental health clinic. It was a very hectic day, full of confusion, misunderstandings, anger, and delays. It was an extremely yang environment. The very next day Gwen phoned to indicate that the fast was a total disaster!!! Monday night she had devoured four dozen chocolate chip cookies. When Gwen was asked what the first thing she wanted to eat was, she said it was a slice of whole grain toast. But she didn't eat it because she had intellectually decided she was on a juice fast. The next thing she wanted was some fruit juice (apple juice), but she didn't drink that either. She drank V-8 juice because it had fewer calories — another intellectual decision. Finally, because of her hectic day and her refusing to give her body what it was asking for, she descended into the cookie jar out of sheer desperation. To slow down, she sought sugar for its depressant, calming down quality. Her body was talking to her all the time, and had she paid attention and had a slice of whole wheat bread or a small glass of apple juice, it is highly unlikely that she would have wound up at the bottom of the cookie barrel.

Body Maintenance

So where does balance come in? Look at those balancing foods in the middle of the yin/yang chart. By eliminating your non-survival thinking that's probably what you'll be desiring. Can you see why it's quite difficult to tell anyone what a balanced diet is for them? Your body is the only one who knows.

A finely tuned machine takes much less fuel and lubrication to run it. So it is with us. The pendulum has a nice easy swing, gentle and smooth, and I believe that is our goal to narrow the swing of the pendulum so that we don't experience such wild extremes. Balance, to me, means experiencing a free-flowing of life, creativity, happiness and energy in abundance, needing less sleep and food. The source of the following is unknown.

> The true basis of the universe is stillness, its real condition, for out of it comes all activity. The ocean, when wind ceases, is calm again as are the trees and grasses. These things return to stillness, their natural way. And this is the principle of meditation. There is night, there is day, when the sun sets there is hush, and then the dead of night when all is still. This is the meditation of nature.

How do we achieve balance? One step is to eat just balancing foods. That might help. But because life is activity, I believe the bigger key is in what we do and how we think. This is why quiet time, time for yourself, is so important during your day. It's the storing up of energy and setting the pattern for the swing of the pendulum at the beginning of the day, mid-day and evenings. If we can learn to listen "by feelings" to what life thinks would be the best way to express ourselves at this moment, we can be involved in fulfilling activities — even if it's spending energy, like tennis. The joy we receive fills us back up. That's how food for the spirit works and it's yin. It adds "sweetness" to our lives. We need to be doing, creating, achieving, giving, moving (yang). We also need to be not doing, letting ideas just germinate, resting, and receiving (yin). Yin is the only place life can talk to us, unfolding intuitive impressions — in our feminine, receiving nature.

3
Removing Excess Baggage

Use three physicians still: first, Dr. Quiet, then Dr. Merry, and lastly, Dr. Diet.

—John Harington:
The Englishman's Doctor, 1608

When the word weight is used in our society it generally denotes overweight, and that's what this chapter is all about. It's not the most pleasant subject, but it is certainly one that gets a lot of attention. For those who have struggled with this problem, it carries negativity, pain, and discouragement. However, let's think about fat in a different way, a fresh new viewpoint often opens new doors. Let's think about fat as a symptom of something out of kilter in our lives, just as ulcers indicate a mishandling of stress. With this attitude, we won't hate fat; we'll merely observe it as an indicator of a correction that needs to be made. And while we're at it, let's also begin to love thinness and start appreciating our bodies.

If you observe thin people, you'll find something interesting about their eating. They don't reason "why" they are or are not eating something. They don't justify overeating; in fact they can't stand the miserable stuffed feeling. This is very hard for a person fighting with weight to understand. Children seem to have this same kind of attitude about eating, except when well meaning parents have interfered with this intuitive process. In fact, children will pick a very balanced diet over a period of two weeks to a month if offered a wide variety of food. So hunger must somehow be an intuitive thing, a perfect guide to the kind and quantity of food necessary to keep us healthy. Animals seem to know when they need certain foods, too. And all of this depends upon enjoying the taste, because when the taste buds of laboratory animals are clipped, the animals no longer pick a balanced diet. Enjoyment is a key to thinness as it is in every other thing we do. Certainly we enjoy quenching thirst, which is definitely an intuitive action, and we depend on that all the time. We don't intellectualize about whether we've had too much water, or how long it's been since we've had a drink. Neither do we always drink water until our stomachs are bulging. We drink only until we are no longer thirsty. That is how hunger operates when we allow it to work for us.

We have reasoned and intellectualized the need for balanced diets and three meals a day until we are the most nutritionally knowledgable and overweight people in the world. This is a scientific fact. You have no idea what the body wants or needs when you desire and eat an apple. The body does know, though, and so it takes what it needs and gets rid of the rest. And what about the fact that foods are produced seasonally? All of this indicates that we haven't given this intuitive, perfectly functioning mechanism a chance to work. We have wanted to trust what books say but not how the body feels.

How does the body tell us what it needs? Through desire. Let's explore that a little. You know the body needs attention in a certain area when it itches. When you are breathing impure air the body almost craves clean air. You feel the need to brush your teeth, often after meals, but certainly when a thin film is present. The excretory needs also make themselves known and you respond accordingly.

With food, your sense of taste is the key—the simple idea that milk would taste delicious, or a spinach salad really sounds great. These things will taste extra good if they are what the body needs. When the need is fulfilled, the food won't be nearly as tasty and we'll no longer desire it, *unless* we are telling ourselves that we should not eat more!

It's been said that there is no such thing as a fat animal out in the wild. If a python snake has eaten a whole antelope, he cannot be persuaded to eat a rabbit, which ordinarily he loves, for days and sometimes weeks. He doesn't need more food and therefore doesn't desire it. If we learned how to listen to our bodies, they would talk to us at a particular meal. What we do, though, is worry ourselves sick over that plentiful meal and fear that we'll gain, hate ourselves for overeating, and eat more to soothe ourselves because we know we shouldn't eat. This really sponsors more hunger and then we gain weight like crazy!

At this point it may be a bit difficult to accept but I am quite sure that losing weight can actually be an enjoyable adventure. I have seen too many people lose pounds aplenty once they began redirecting their minds toward fun, simplicity, and excitement.

You will begin to develop your awareness of being able to listen to your body and place significance of events that happen in your life by learning to become still, listening to a silence within your own head and body. Quietness in your life is important: it is a time for thought and easiness. Just notice the beauty of your surroundings. Make your personal world special and establish a feeling of happy expectancy about your day. Say something like, "I wonder what good and exciting things await me today?" Feel like all is well between you and the world, and then look for all the usual and unusual things to happen that will be supporting your new beliefs about you. When you are constantly in a relaxed state, you will be able to see how you add newness and growth day by day to your life. Carry a small notebook and write down the interesting new insights you have. Record such observations as, "I started to eat something and noticed that it coated my tongue or tasted peculiar." Awareness begets more awareness, just like success breeds more success. Like attracts like, and this is the big key to being able to become slender and stay that way.

Can You Imagine That?

The mind controls the body. Even though this has been known for thousands of years, we have not made good use of this knowledge until recent times. Edmund Jacobson proved that when you think or imagine something is happening in your body, it in fact is happening. This is being confirmed all the time now in biofeedback and you can use your imagination to lose weight.

What is the imagination? The mind works on pictures and feelings, not words. Dr. Maxwell Maltz said that imagination is the real powerhouse of the mind and that any time there is a contest between imagination and will power, imagination always wins. For instance, if you are trying to *not eat* a certain food, you are thinking about it in pictures and feelings or imagining it. The imagination controls the willpower and you will probably end up eating that food. We've been told to use more willpower, but that is like trying to keep the muscles of the body flexed indefinitely. It can't be done, so we have been fighting a losing battle when we diet.

Humanistic psychologist Abraham Maslow said that whenever we try to suppress a basic need, and hunger is one of them, we only become possessed by it or obsessed with it. Again, that is saying that the imagination is brought into play against us whenever we try *not* to do something. This is why diets fail. Just the word "diet" often carries a lot of negative feelings about being deprived and hungry. This is why we end up going on binges. Willpower is meant to be used for decision-making and then only on a short-term basis. To effectively use the imagination, imagine what you *do* want (a thin body), not what you don't.

Begin to imagine what you want, thinness, and desire that with a feeling that is yours alone. Imagine how delightful it would be and let the imagination play with that idea. Develop a curiosity about thinness.

What would that be like, to be permanently thin? What would people say to you? How could you feel?

It has been proven that the mind cannot tell the difference between something really seen and something imagined vividly and in detail. So, Maxwell Maltz says to think of your goal or end results in terms of a present possibility and make it so real to your brain and nervous system that you have the same feelings you would have if you had that result right now. How would you feel if you had thinness right now? Wouldn't the feeling be outrageous delight? Joy beyond description? Remember, the mind works on pictures and feelings, so capture that feeling of joy of having thinness as often as possible.

You've probably heard the statement, "It's done unto you as you believe." Certainly if your mind believes whatever it sees, whether with your eyes or imagination, you can use this to get thin. David Seabury said anything you tell yourself over and over again in a convincing tone of voice, you will believe. Napoleon Hill calls this autosuggestion. These statements or affirmations make an impression on the subconscious mind and the subconscious mind is the "doer" of the mind, the part that gets things done. It acts on your believed thoughts or feelings and sponsors actions that are normal, natural, and automatic. Tell a child that he is stupid often enough and he will be able to perform only in a stupid way.

What's Your Expectancy?

A study was done with a large group of children regarding expectancy. A large number of them were taken aside and given a standard IQ test. The IQs were recorded in the IQ column on a chart. Those children with high IQs were told what it meant and what to expect, and their teachers were also informed. The same was done with those with average and low IQs.

Another group of children was given the same test, but instead of their IQs their locker numbers were listed in the IQ column , and the children were told the locker number was their IQ. Those with high locker numbers were told what it meant and told what to expect, and their teachers were likewise informed. The process was repeated for those with average and low locker numbers. One year passed, and upon examination it was determined that those children with high IQs were high achievers, those with average IQs were average achievers, and those with low IQs were low achievers—just what you would expect. *But*, those children with high locker numbers were high achievers, those with average locker numbers were average achievers and those with low locker numbers were underachievers and failing in significantly high proportions. The reason? Expectancy. If you expect to be a dummy, you will be a dummy.

Say something that you desire over and over to yourself in a convincing tone of voice, and your mind has to accept it as true. How about, "I'm thinner today that I was yesterday, and I'm so grateful." Gratitude only helps us to be better receivers; we can't be grateful about something we don't have, so use it to help convince your mind that you already have this thing. In this case you are both giver and receiver—if you have trouble receiving, you'll have a small problem here. You are giving something nice to your body; can you accept it? Can you accept a compliment?

You will begin to notice striking changes. The biggest changes will occur in your appetite, in both kind and quantity of food you desire. You'll seek more activity and maybe less sleep. You won't seem to notice the clock as much. All of these things seem to come as normal, natural, and automatic actions, and are nothing you have to force at all.

In a study at the University of Pennsylvania a group of overweight people were tested on their desire for food in relationship to the time of day. Three groups of people were situated in three different rooms. Each room contained a clock. The clock in the first room was set to run slow, indicating that only fifteen minutes had elapsed in each actual half-hour period. In the second room the clock ran normally, and in the third room the clock ran fast, indicating that one hour had elapsed in each actual half-hour period.

All the people under study were asked to fill out various papers (just to keep them busy). The people were allowed one actual half hour to do this. Examiners then went into each room, bringing with them sandwiches and snacks. The people in the first room thought only fifteen minutes had passed, and they ate some of the sandwiches. The people in the second room thought one half hour had passed, and they ate more than the people in the first room. The people in the third room consumed all the sandwiches, for they thought an hour had passed. The study indicates that our eating habits and desires for food are often governed by the clock. At noon ask an overweight person what time it is and this is what you'll probably hear, "It s lunch time." Your awareness is a key to becoming thin.

Right before falling asleep is a very good time to establish the feeling of having thinness. Try to say it until you feel warm and excited about it, and maybe even have goose pimples or tears in your eyes. Sometimes cutting a picture out of a magazine of someone who looks like you would like to look will help your mind get the picture. For goodness sake don't pick out a girl in a bikini or a Mr. America if you can't see yourself that way or just don't like it. Make it as close to real for yourself as possible. Put it where you can see it often and think of the joy of being thin.

Tell yourself that you are going to depend upon your hunger to guide you to thinness and keep you there. Get into the spirit of childlike fun by talking to your body. You may want to tell your body that you know you caused it to be in the condition it's in, but be persistent, because you have a whole new blueprint for your body to follow and it's going to feel so much better and have to work less in no time at all. At first you may feel a little silly, but the more fun and light-heartedness you can put into this the more you'll enjoy it. All this loving concern for your body is probably new to you, since most fat people do not love their bodies.

All life responds to praise, appreciation, love, sincerity, kindness and good manners, so use all of these on yourself and your body. Nothing responds in a positive way to condemnation: "bless a thing and it blesses you; curse a thing and it curses you back." Know that you're in complete charge of this exciting project; you don't need the support of anyone outside yourself. You must become your own best friend, your biggest fan club, your most loyal supporter!

Whenever you are hungry, sit or stand, take a couple of deep breaths, let your arms hang loose and say, "What do you want to eat, body?" This eases your attitude, makes you receptive, gives you a concerned feeling toward your body, and relaxes you. Eat only when you are hungry and don't worry if you feel hungry at a different time from the three "regular" times. You may really be hungry then, but your activity will dictate much about the food you need. Really pay attention to this hunger, but also pay attention to when you don't want any more. Deliberately notice and enjoy the taste of every bite you put in your mouth, because when it begins to lose its super good taste, your desire will diminish too. You will be surprised at how often you will only want a few bites of something and not even get close to that *full* feeling.

Sometimes you will feel hungry, but not for physical food. Man can't live by bread alone, so learn to distinguish between the two. Mental food or food for the spirit comes in many forms — love, music, knowledge, exercise, beauty, laughter; something as simple as saying kind words to a loved one in a strained situation can help. On the more mundane side, balancing your checkbook or some other chore can provide a surprising relief and a kind of nourishment. You'll learn by practice what hunger is telling you, especially if you eat and are still hungry!

Don't have a list of "never eat" foods; that's the perfect way to sponsor cravings and binges. Just let your hunger guide you. If you really listen, it will never steer you wrong. If you have a fear of a certain food at the present time, tell yourself you're going to protect yourself by substituting something else that would not frighten you at this time. This is only if you happen to "think" of that food and have a feeling of worry about it, or feel that it is really fattening and will stick to your body in the wrong places! Some people actually think one little cupcake will put five pounds on them. You will eventually lose this fear of a particular food, but in the beginning consider the fear natural, since it is the way you have been believing. What usually happens is that a binge begins from the one indulgence; people binge on Mexican food from just one bite, or perhaps chocolate has a particular power over them. Just be calm about this and try to gently let go. You may be able to get so involved in fixing a beautiful creative salad that you forget all about Mexican food—there is an interesting feeling that goes with doing something special "just for yourself." Try it.

If you should stumble a little by eating more than you really desired, forgive yourself, forget it, and treat the new moment as a new beginning. Tell your body to take what it needs out of what you have just swallowed and get rid of the rest and start all new and fresh. Try as much as possible to not think of foods as either "good" or "bad"; allow your intuition to work for you, even if it doesn't seem logical to want beans three times in a row.

Beans Away!

Glynis Johns, the actress, was on a tour through Europe. She had a desire to eat beans three days in a row. One of her friends told her she should eat more of a balanced diet instead of all those beans, but she followed her desire and ate them. Later she discovered that beans contain the B vitamin, which is good for nerves — just what she needed. If we just follow our desires we might find a whole new life out there.

I have found that sometimes fasting is an extremely valuable tool, if I prepare my mind first. What you put into your mouth and your stomach does affect your attitude. When I'm low in spirits I sometimes fast. Fasting from a sense of feeling deprived is devastating, but fasting from a sense of helping yourself to a higher frame of mind is incredibly valuable and uplifting. Just tell yourself you'd like to try to drink only tea, water and juices (I recommend only juice fasts, not total abstinence, or water fasts), but if you feel the need to eat something, that's okay too. When you begin to get a little hungry, sip some juice or hot tea and the hunger will go away. The fasting I do is not to lose weight but to help maintain or get back a happy state of mind. This is only a suggestion; if there are medical reasons why you should not fast or if it gives you a negative feeling, don't try it. But be open to fasting as a possibility in the future. The fasting method has convinced me that a person could lose weight as fast as he or she wanted to without suffering — but realize that it is only an alternative route to add variety to intuitive eating.

Since the body responds to what you are thinking, experiment with visualizing. Imagine your muscles having teeth and eating little bites of fat as you run up the stairs or walk down the street. Picture your body pulling fat out of storage in the places you would like it gone and putting it to very good use or moving it to places you would like more! Think about swimming 100 lengths a day, vividly and in detail. Tell your body you can just see the fat melting off in record time or that you have a special banquet of tummy or thigh fat available today. Never hate the fat; just love thinness. Go look at smaller sized clothing and picture yourself in it. Stand nude in front of the mirror and notice how much smaller a roll is now than last week and then take the time to be grateful about it. Notice how much easier it is to button or zip your clothes and be happy about it—but don't weigh! Hide those scales! You are learning to depend on inner senses, and the scales can destroy in an instant any sense of accomplishment. Can you remember being on a diet, doing so faithfully well, and then getting on the scales only to learn that you've lost only half a pound? You immediately feel discouraged and may want to eat. You don't need the scales to tell you you've gained, and you don't need the scales to tell you you've lost. Fat people are obsessed about numbers on the scales; thin people never or rarely weigh.

Since you are showing kindness and concern for your body, take time to do some physical things to help make this feeling grow. Every day look for something about you that you like and mention it to yourself. Deliberately look in the mirror. When you bathe, take a bubble bath and then dry yourself gently and put lotion all over your entire body with loving care. Wear special underwear and put on your best clothes. Wear perfume or cologne, or aftershave lotion if you like it. Experiment with makeup and hair styles. All of this will help you feel like you deserve everything good, including thinness.

Now the things we have been talking about are all helpful, but the thing that must transpire in your mind and life is the feeling that you are an extremely worthy, deserving person. Having self-esteem does not mean being conceited. You must feel that you are taking your place in life without being better or worse than anyone else. You don't need to imitate anyone; you have a whole truckload of fantastic qualities put together in a unique way. Life thinks you are really special, and you are—but the one person who needs to know and feel it is you. In all time, in all the universe, there's never been and there will never be another you. The way you feel toward yourself is the way you greet the world, and the world responds back just like a mirror. Troublesome people fade away as the result of raising self-esteem. We become more accepting of ourselves and thus more accepting of other people.

One way of working on self-esteem is with affirmation. Say to yourself, usually before you go to sleep and sometimes in the car or around the house, "I am no mistake. Life had a need and a desire to express itself as me, and was so sure the job could get done through me that it threw away the mold and didn't even bother to make two of me. I'm the only me there is and I'm here absolutely on purpose to live, express and enjoy life. Life feels I am special."

You can make up your own little statement full of feeling to yourself; in fact, the more creative you make it the easier it is to feel a loving feeling toward you. You might just want to say you are qualities of life that you would like to display. Remember, the subconscious mind believes whatever you tell it over and over and you will find yourself acting accordingly. So some simple affirmations are: "I am beauty, I am strength, I am joy, I am love, I am warmth, I am receptivity, I am tenderness, I am responsiveness, I am flexibility, I am perfect man (or woman) form." Desire, expectancy and persistence are the keys to thinness. If you desire it, if you expect it, you will naturally be persistent.

4 Health Maintenance

We don't die, we kill ourselves.
—Sinclair Lewis

I am not a medical doctor and I don't intend to get involved with medical terminology regarding diagnosis and recommended cures. In fact, my intent is not to diagnose or recommend cures for anything. As a living, moving, active being on this planet, I have had various illnesses in the past; I have also been quite healthy. Right now I feel fit as a fiddle, in tune with life, alert, and quite relaxed.

I have had many opportunities, as most of you have had I'm sure, to observe my fellow man and myself during these two states, illness and health, and like researchers in the medical profession I have noticed a series of somewhat striking patterns which you may find curious.

My physical condition seems to reflect my trend of thinking of the past few days, weeks or months (the incubation period varies). If the body is totally healthy it seems that one's thinking must have been correspondingly healthy for a time prior. However, if the body is diseased in some way, I have noted that one's thinking a time prior to the illness often has included anger, fear, worry, jealousy, and other similar negative attitudes.

My question then is this: could it be possible for one's mental attitude to be the governing factor in the state of the physical body, be it illness or health? Beyond that, does the individual have the capability of controlling this to such a degree so as to successfully influence the body through the will to live or the will to die?

In past years it has been quite common, in determining the cause of an illness, to study the illness itself, searching for the physical reasons for the illness. A more recent approach is to study the person as he or she moves from disease to health by either normal or artificial means. It seems to me that when the body is in a state of disease it is experiencing exactly that — a state of dis-ease. If the body is uncomfortable, irritated, or under stress, do you suppose that putting it into a state of ease or relaxation might help things along a little?

A Warm Hand Please

It's been known for some time that the yogi masters are able to control their bodily functions to such a degree as to even stop the heart for periods of time. Recent studies into the biological functions of the body have produced instruments for instant readouts of many of these functions, such as body temperature, blood pressure, heart beat, brain wave patterns, and the like. This readout process is called biofeedback.

With constant observation of the body processes, by monitoring the biofeedback device, a person can learn to influence body functions that have been previously considered strictly involuntary. This can be done by monitoring the instrument for heart beat, then counting faster or slower to raise or lower it. Or in the case of body temerature, simply imagining the hand is in a warm bath of water will cause the hand temperature to rise. Increased blood flow to any part of the body can be produced in the same manner.

If this is possible, then would it not also be possible to influence the activity of the body's natural immune system and direct an attack on a specific diseased area of the body? What would be the key to such an attack?

My experience is that it *is* possible and the key to such a process lies in the magnetic patterns of mental imagery and expectations. It's been said that there are no victims in life, just volunteers. The big problem has been that we haven't understood the important role of thinking. We have become ill and said, "Well, that's life. You've got to take the good with the bad. The flu's going around." This attitude holds that we are just victims of circumstance—but we are not! Notice that when the newspaper reports that the flu season is here the number of flu cases more than doubles that week. Some people hang on to the seven-day cold for seven days or longer and some kick it in one day or never get it at all. Studies often show that a placebo (sugar pill) can be used as successfully as an actual drug.

In all these cases it has been determined that the psychological makeup and the general patterns of thinking of the individual play an extremely important role in the susceptibility to disease and recovery from it.

Stress Is a Myth

There is no stress in the universe. There is no stressful situation. You can't take the stress out of the air and put it in your stomach. You *can* take ulcers out of your stomach. Ulcers are a mishandling of "stressful situations," of which there aren't any. You see, we choose to look at something as stressful. We become wrapped up in it emotionally, then choose to let it affect us. In other words, we think stressful thoughts. That's the only kind of stress available to us.

Now, remove the ulcer. If you are still thinking stressful thoughts your ulcer will return. Remove the ulcer again. If you are *still* thinking stressful thoughts your ulcer will return once more—and on and on. You see, the ulcer is only the effect of your stressful thoughts. If you really want to get rid of the ulcer for good, you surely must change the thoughts in your head. To think you can cure the ulcer completely by the knife is sheer folly.

What seems to happen is that by maintaining a stressful attitude the body's response regarding the immune system decreases, thereby not providing the proper support necessary to sustain health. If people would support health as much as they support sickness we would all be much better off. Bathroom medicine cabinets are filled with aspirin, sleeping pills, bandages, iodine. Ask yourself what you are preparing for. It seems that the commonly accepted method to "solve" anything these days is to take a pill — aspirin, tranquilizers, antidepressants, sleeping pills. In one year alone over 50 billion aspirin tablets were sold. That is equal to one tablet every day for everyone in this country over the age of 5 years, except on weekends.

We have come to rely on a pill to solve our problems, but problems there will always be, pill or no pill. Life is a series of problems—big, small, fat, short—and it's not the problem that's the problem; it's how you view the problem. Your attitude is the key and your understanding is the answer.

Keep Your Car in Tune

Most people perform preventive maintenance on their car. Those that do change the oil about every three or four thousand miles, change the oil filter, get a tuneup every now and then, rotate the tires and generally maintain the automobile in such a manner so as to minimize the chance that problems will occur at inappropriate times. If people would spend as much time and thought on the vehicle that walks them around as they do on the vehicle that drives them around it's a sure bet they'd be in much better shape. When your car is in desperate need of oil you don't drive it for another three or four hundred miles before adding the oil. You know that if you don't fill the tank with gas when it's low you won't be going much farther. When the tires get bald it would be senseless to begin a long journey without getting some tread under you. The body, on the other hand, can be crying out for a certain kind of food which is withheld because we are too busy to eat, or are trying to lose weight fast. It can be in desperate need of rest or even sleep but we say we've got too much to do and can't stop now. Yet the body is saying that if we don't listen to these messages, it will make us hear by giving us louder ones.

The louder message may come, for example, in the form of a sore throat. The body says, "Can you hear this?"But if you don't give it much thought and go to work anyway, the body will start to speak in a little louder voice—you develop a cold and possibly a cough. The voice will get louder and louder until finally you'll have to listen, thereby changing your thinking and working patterns to accommodate the illness with corrective measures.

We don't often see a sick animal. One reason is that animals don't worry, and they certainly don't push themselves. They live in the now, expressing each new moment. On occasion, however, animals do become ill. One reason is that they take on characteristics of their owners. Now, what does an animal do when it is ill? Generally it sleeps a lot. It assists the healing process by obeying the natural laws of recovery. A dog doesn't play just one more game of catch the stick because it hasn't made its quota for the day. No, it rests.

A Day Off

Imagine walking into work some morning, going directly up to the boss an saying, "Boss, I feel great today, I'm taking the day off." He'd think you're crazy. People don't take off from work when they feel great; they only do that when they're sick. As far as I'm concerned, sick leave is just another excuse to get sick and we've got enough of those already. Besides who wants to waste a day off from work in bed?

A woman I know gives her kids "mental health days" during the school year. When they feel like a day off or want to do something special, they don't have to get sick to do it. They just ask for a mental health day. Perhaps because the choice is there, they don't even take advantage of it. It's possible your kids wouldn't either, nor would your employees. A little consideration and compensation for feeling good and doing a great job can go a long way. Try it sometime.

There is a process that I have used, with much success I might add, in moving the body back into a state of health. For the moment don't condemn yourself by fretting or worrying about how you got sick; just accept the fact that this is where you are and take it from there. I hope this point is well made, for it is most important that you in no way condemn yourself for your present state. I have found that any thoughts of condemnation, guilt, or anger for the state of illness you may find yourself in definitely impede the healing process.

It is absolutely imperative that you change your basic trend of thought. Whatever situation that seems to be foremost on your mind, whether you think it has anything to do with your illness or not, *change it.* Or better, don't think about it at all. Tough, right? But *it is imperative that you change your basic trend of thought.*

How do you do this? Try the following little exercise.

Imagine a dog lying in the road, just after being hit by an automobile. It's a little puppy, and it's hurt.

There's some blood and it's whimpering a little. That's not a very comfortable visual imagery, is it? You may feel sympathetic and compassionate for the puppy; you might even feel sick at your stomach.

Now switch your attention to a rose garden. It is outside and the roses have a fragrance that fills the air. Reach down and smell an individual rose. Feel the softness of its petals—almost a silky smooth feeling. What color is your rose? Be careful of the thorns on the stem for they are quite sharp; handle them gently. I love roses; I feel good when they're around.

What happened to the puppy? You see, when you think of a rose, you can't think of a dog. When you think of something beautiful, you can't think of something hurtful. You *can* change your thoughts. It takes constant watching, for a trend of thinking will struggle to maintain its own existence. I have found that it is possible only to observe the thoughts as they go by, not to plan what future thoughts will be, and not say that the one I just thought was wrong. Now, this present moment, a rose is a rose. This moment is all you can take care of, especially in unfortunate circumstances — like your spouse walking out, deteriorating health, or financial collapse. When you are redirecting your thoughts, do this with a relaxed body. Loosen your muscles; ease up a bit. This will arrest the declining health; it will stop it in its tracks. But we want more than that. We want a healthy body and we want it now. So what to do?

What You See Is What You Get

The old saying is especially true when dealing with the mind. But it could also be stated: "What you see is what you become." Remember that experimental and clinical psychologists have proven that the mind cannot tell the difference between an actual physical visual impression and one imagined, vividly and in detail. Since we all think in pictures and feelings, this part will be quite simple, for you'll be picturing as we go along.

Understanding reduces the complex to the simple. By understanding the workings of your own mental camera you'll be able to focus in on your desires and bring them into your own physical reality.

There is a certain basic way in which we think of ourselves. The pattern of thought or belief that you have about yourself is carried with you and colors your experience.

Three Moves to Failure

I know a man who, for a long while, lived in New York. Things were going quite badly for him. Money wasn't coming in, his wife turned against him, his friends seemed to drift away, and in general he was having a bad time of it. He got the idea to move. After giving it some thought, he packed up and went to Chicago, looking for that fresh start, that new beginning, the new life he'd heard about.

Chicago offered new challenges, and he met them — in much the same way as in New York. His financial situation did not improve, he seemed to make only acquaintances, not friends, and in short, for all he knew he could have been living back in New York.

Thinking that a warmer climate would solve his problems, off he went to Los Angeles: balmy breezes, beaches, sun and fun. He met a fine young woman, established a business, and the pattern began over again. This woman, too, left him and the business was another financial disaster. Why? Because wherever he went he took himself with him! When you want to change anything, from health, to business, relationships, finances or whatever, you've got to leave the old you behind and move forward in the new direction with a new attitude. You've got to *see* yourself in your mind as being a different person. That's where your mental camera comes in.

Crystal Clear

What I'm talking about here is forming imagery patterns of your desire in your mind, crystalizing an image on your mind, taking a mental photograph of those things yet unseen in the physical world.

Thinking is the beginning of forming an image on the film of your mind. A really sharp image is ideal, but acquiring one may take some practice. We all imprint our minds with imagery of amazing clarity in our dreams and our fantasies. This process is called hypnogogic imagery, or creative imagery. It is a state that the mind travels through just prior to falling off to sleep. Once you learn to relax the body completely without falling asleep you will be able to experience this. You will see clearly patterns, people, scenes, usually common ordinary images, not at all intertwined with the symbols which often occur in dreams. By remaining unemotional when these images come into your mind you can retain them and experience this vivid imagery as a comparison to your own predetermined mental pictures.

Feeling Is Important

Imagery is a valuable tool to develop a feeling. The body responds to feelings as well as to pictures, and if you really have a gut feeling of health, you will be healthy. If you don't have an extremely strong feeling of good health, through mental imagery a definite feeling can be developed.

In a sense I guess you could say that you're tuning into your inner senses. What we want is something we don't have, and to get what we don't have we have to use what isn't. To achieve health, wealth, happiness we are again turning to the physically unseen, using a law that can be demonstrated by anyone by direct application. By clicking the shutter of your mind and forming a series of mental pictures of your desire, you set up a harmonious vibration with that desire and set yourself to attract it as reality.

A Lovable Little Seed

In my seminars I carry around with me a seed, usually an avocado seed. Now some of these seeds I plant in my pottery, and when they grow to a reasonable size I give them away as gifts. It takes time to grow an avocado tree, as well as plenty of sunlight, water, and proper attention. The same is true with a little baby. It needs attention, watering, feeding, changing, love, warmth, holding, compassion, understanding, and more. A child tossed out in the cold and forgotten can't be expected to grow; it takes time and nourishing to "grow" a child.

Now we get to our mental seed. We plant it in the mind by forming a picture of it, and because we don't see immediate results we condemn it and say, "This doesn't work," and we throw it aside in disgust.

Transmutation

You cannot escape the results of your thoughts. You are not what you want; you are what you think. You will realize the *vision* in you mind, not the idle wish. A vision held frequently and intently will tend to demonstrate itself more quickly than one held in the mind only on occasion. Frequency and intensity seem to be very important in making your dream a reality. Just as the seed planted in the ground requires time to grow, so does a vision need a gestation or incubation period.

There is no set time period necessary to manifest an idea. However, we do know that the speed with which change occurs is directly related to the frequency with which one thinks about it and one's emotional involvement with it. Frequency and intensity are two key factors in bringing your true heart's desire into reality. Most people plant the seed and then fluctuate in their response and intensity. From expectancy (looking for results in the first hour, day, or week, depending on the person) and excitement, they swing to doubt, hope, or wishing. From sureness, firmness, and affirmation, they switch to concern, despair, and uncertainty. Then follows a period of re-excitement, interest, and enthusiasm, perhaps succeeded by a new period of fear, anger, and depression.

And so it goes, back and forth. More often than not we end up not getting our heart's desire, whatever it is, because we vascilate too much. It's not because the idea wasn't any good; remember, the mind doesn't discriminate. It's because we have bounced back and forth between confidence and defeat. It's as if the universe is saying, "Hey, make up your mind so I can give to you." For every contradictory thought we hold we extend the time necessary for the desired change to become apparent. By planting weeds along with our crop we get a jumbled mess, and sometimes we can't even recognize the sprout when it appears. The farmer doesn't go out each day to dig up the seed to see if it has grown!

Passive or Aggressive

Magnetic imagery will always attract an environment to you, but we're also interested in speed. We want to get this thing taken care of—either get it into or out of our lives as quickly as possible. Because of this the method in which you form your mental pictures becomes extremely important.

Picture the end results in terms of a present possibility. For example, let's say that you are told there will be a possible delay from a business trip, and you don't want to be home late. A procedure to follow could be picturing yourself returning home, greeting the family and turning on the 5 o'clock news, confirming that you arrived home safely and on time. This is called passive imagery, and it will certainly be effective. However, it's not a very dynamic mental picture, and although it's possible to frequently picture this in your mind you may have a little trouble getting the necessary intensity behind it to bring it into reality.

Common (Unaware) yo-yo type thinking

A. Plant seed (Expectancy)

first hour, day or week
.... depending on person

B.

Second hour, day
or week.

Excitement

Doubt
Hope
Wish

Sureness
Firmness
Affirmative

Concern
Despair
Unsure

Re-excitement
Interest
Enthusiasm

Fear
Anger
Depression

Now, take the same situation of possibly being late returning home, keeping in mind you really want to be there. You begin picturing this: As you open the door to your home you are hit by the tantalizing aroma of roasted duck a l'orange. You close the door behind you and the old grandfather's clock right in front of you strikes five. The room is dimly lit by candles; the table is set for two. As your wife turns *off* the 5 o'clock news she comes over to you, puts her arms around you and says, "I'm so glad you're home on time, honey; I have planned a special dinner just for us tonight." Can you see the difference? This last example would be more in line with aggressive visualization.

Both passive and aggressive visualization can be effective, but aggressive imagery brings into play all the senses—sight, sound, taste, smell and touch. The "on time" sequence was hit three times—by the clock, by the TV and by the wife. It is possible to achieve both frequency and intensity with the aggressive mental picture.

Another example of visualization might be this. Imagine that you want an area cleared and a hole dug for a swimming pool. You have men working on it, but progress is slow. You may be picturing the men doing their jobs, but in a very passive and lazy way, thereby causing the job to continue at a slow pace. Sure, it will get completed eventually, but you want it done fast. Try aggressive imagery. Picture bulldozers with huge scoop shovels throwing dirt every which way. Men are running to keep up with the pace. There is loud noise from the machines, and black puffs of diesel smoke are present. The men shout back and forth, calling for more concrete, *fast!* Imagine yourself out there on the dirt mound. See the loose dirt piling up; smell it, feel it, even taste it if you want to. Get yourself into the picture and live it as a present moment. Get it into your mind that this is happening now, not in the future.

Could you use the same technique with a bodily disease? How about cancer? If a person were working on attitude to control the disease he could form pictures of the cancer as a granite wall, the white blood cells attacking the cancer as little men with small picks and shovels. This would be most definitely passive imagery. Now, take the same person and have him imagine the cancer as little black peas. The white blood cells are thousands of wolverines, devouring the tiny black peas in their enormous hunger. Can you see the difference? The second is aggressive imagery. If visual imagery is effective at all, the latter would certainly be a far superior mental image to form.

Another example might be for the person to see the cancer as a huge mound of concrete, with their treatment (such as chemotherapy) as a river, gradually eroding the mound. The white blood cells are small scrub brushes that come and clean up around the edges. Clearly, this type of imagery doesn't indicate a belief in the possibility of overcoming the cancer very rapidly. A more aggressive picture would be to see the cancer as weak and easily overpowered by the treatment and natural body immunity, such as by picturing the cancer as hamburger being devoured by a huge army of polar bears. (Carl O. Simonton, M.D. relates this example in *Cosmopolitan Magazine*, Summer 1978.) The law of polarity says that like attracts like. What are you looking at—the positive or negative side of life? I cannot stress enough that what you think about you draw to you like a magnet. If change is to come you must grab hold of a brand new idea and use it!

Anything You Want

From the sound of all this it certainly seems as though we could have anything in life that we wanted. After all, we know that people under hypnotic suggestion can get rid of their inhibitions quite fast, enabling them to exhibit extreme confidence in front of an audience, eliminate stuttering, even improve their sports activities. Why then can't we, in our normal conscious state, have anything we desire?

Let's face it. Some things come easier than others, and it all has to do with our true desires. If sports come easier to you than playing the piano, if you really like sports, you'll probably find yourself out on the tennis courts much more often than sitting at the piano practicing the Rachmaninoff concerto in C# minor. Playing the piano would be nice, but you *love* sports. See the difference?

I believe that life speaks to us through desire. De-sire means "from the source." If you truly listen to your desires, in any area, you can rest assured that if you follow them you'll be on the right track.

Fly Me to the Moon!

I went to talk to some people one day at a public school. After my talk one of the men came up to me and said, "I think what you say is ridiculous. You can't have anything you want. I've always wanted to go to the moon, and as you can see I don't have what I want!" I said, "You don't want to go to the moon."

"What do you mean?" he asked. "It's been on my mind ever since I was a kid." "Sorry," I said, "I can't buy that, you don't want to go to the moon." I asked him if he'd ever taken any flying lessons, any courses in aerodynamics or aircraft flight systems? Had he ever studied any meteorology, physiological functions in zero gravity, or advanced mathematics?

"Well, no," he said.

"Have you ever applied to the Houston Space Center for a position as an astronaut?"

"No," again.

"You don't want to go to the moon," I told him.

"I see what you mean."

I asked him what he really liked to do and he said he liked teaching kids. That's exactly what he was doing. I said to him, "I'll bet you never felt about the moon like you do about seeing kids learn."

"You're right," he said, "it gives me such a thrill when they grab on to a new idea and begin to apply it in their lives. I love it."

Yes, you truly can have anything you want. You don't have to take what you get; you can get what you want, but first and foremost you must be true to yourself. What is *your* heart's desire? Working toward anything else is misdirected effort.

The Pay-Off

You might have asked yourself why some people have such junky lives. Usually it's because they do not know a better way. They choose junky thoughts and for the present they are satisfied with the pay-off—sympathy, self-pity, and other non-survival attitudes and reactions. It's like going to a movie; you make the choice, you pay your money, and you go in and see it. The film may be humorous or it may be a horror film. You know that you can get up and leave any time you choose, because it's only a movie. Why would you sit there if you didn't enjoy it? For no reason at all. Still, we sit through life, flashing illusions on our minds that we don't enjoy all the time because we have not had the awareness, self-discipline, or understanding to do things differently. There is only one reality, and that is *yours in this present moment.* You choose it; no one else can live it. No one else can even see it as you do, for no one else can get in your head to look out from within. It is a sense of aloneness that everyone experiences. You are the only person who can perceive the world through you, and you are the only person who can run your projector. We too often attempt to beat up the screen when things aren't going right; we blame outside circumstances and point fingers at *them.* Well, life is just like the movie in the theater. If you don't like what's on the screen the only way to change it is to run a different film in the projector. Your mental projector can do that for you!

PART TWO
business

5
The Universal Force Field

The quick harvest of applied science is the usable process, the medicine, the machine. The shy fruit of pure science is understanding.

—Lincoln Barnett: "The Meaning of Einstein's New Theory"
Life Magazine, January 9, 1950

"Everything moves, everything vibrates, nothing rests." This was recorded as being said thousands of years ago by Hermes Trismegistus in ancient Egypt. Today every first-year physics student knows that everything moves, everything vibrates, nothing rests, but how did Hermes know that way back then? Is the atomic structure of the universe a relatively new discovery, as most people believe, or it is one of the age-old ideas of the past? We may never know, but this we do know that locked up inside that statement is the key to understanding the patterns of life that we deal with in this, today's world.

With understanding of the above statement, I believe it is possible for an individual to achieve total health, wealth, and happiness, right here, right now. It all has to do with vibration.

Let's begin by investigating what is meant by "everything moves." Imagine that you have in your hand a common, ordinary ballpoint pen. Let's say it has a metal casing and is silver in color. Physics tells us that the actual structure of that pen is made up of extremely fine particles called atoms, composed of electrons, protons and neutrons, all moving about at such extreme rates of speed much too rapid for the human eye to see. Thus the metal appears to be motionless. Further investigation into the structure of this pen will reveal that there is actually more "space" between the atoms in relation to the size of the pen than there are atoms. So we have determined that there is more space, or "nothing," in the pen than actual solid matter.

The question now pops up as to why it looks solid. It looks solid because the particles are so small and so closely interwoven together that they actually appear to be touching one another. This of course is a visual observation made with the human eye only. Under the electron microscope it is quite another story.

An idea of the smallness of the particles can be gained by this analogy: research has indicated that if everything on the planet — earth, food, cans, cars, people, clothes, and the entire planet itself—were to be condensed together, without any space or air between any of the atoms, we would have a mass about the size of a baseball. It would be heavy, though, and we would probably have a hard time getting it out to left field.

You see, our visual sense only perceives the pen as mass. What it is in reality is very little mass and a whole lot of electromagnetic energy, pulsating, vibrating and moving right in your hand as you hold it. Like the pen and all matter, so is the Earth. We are really then walking on an illusion, walking mostly on space or an energy field.

Now, we perceive the metal casing of the pen as metal because various atoms have hooked up with other atoms to form molecules—a molecule being the smallest particle of something that still is the thing. However, behind that particle there is an even finer particle. It may not be metal, because that combination of atoms may not combine as metal, but it is a finer particle. And behind that one there is a finer particle, and behind that one there is another one, and so on. And as long as you have a "that one," there will always be something behind it. Everything we can label as finite in the universe, anything we can conceptualize, is made of something finer than itself.

At a place where that which is finest, so fine that the tiniest parts of atoms are made up of themselves, there is a place where everything is the same. There is no more uniqueness at this level; it's all homogenous. The universe in a sense is solid. You see then that an apple, plant, your head, candy, air, a candle and the flame, even a vacuum are all the same.

Vibration

Okay, back to the pen. Let's take that same small ballpoint pen and attach it firmly to a vibrating machine, a machine which has infinite speeds of vibration. We turn on the machine and the pen begins to vibrate slowly, back and forth. We increase the speed a little and the pen moves faster. A little more and soon the pen is moving so fast we can't see it any

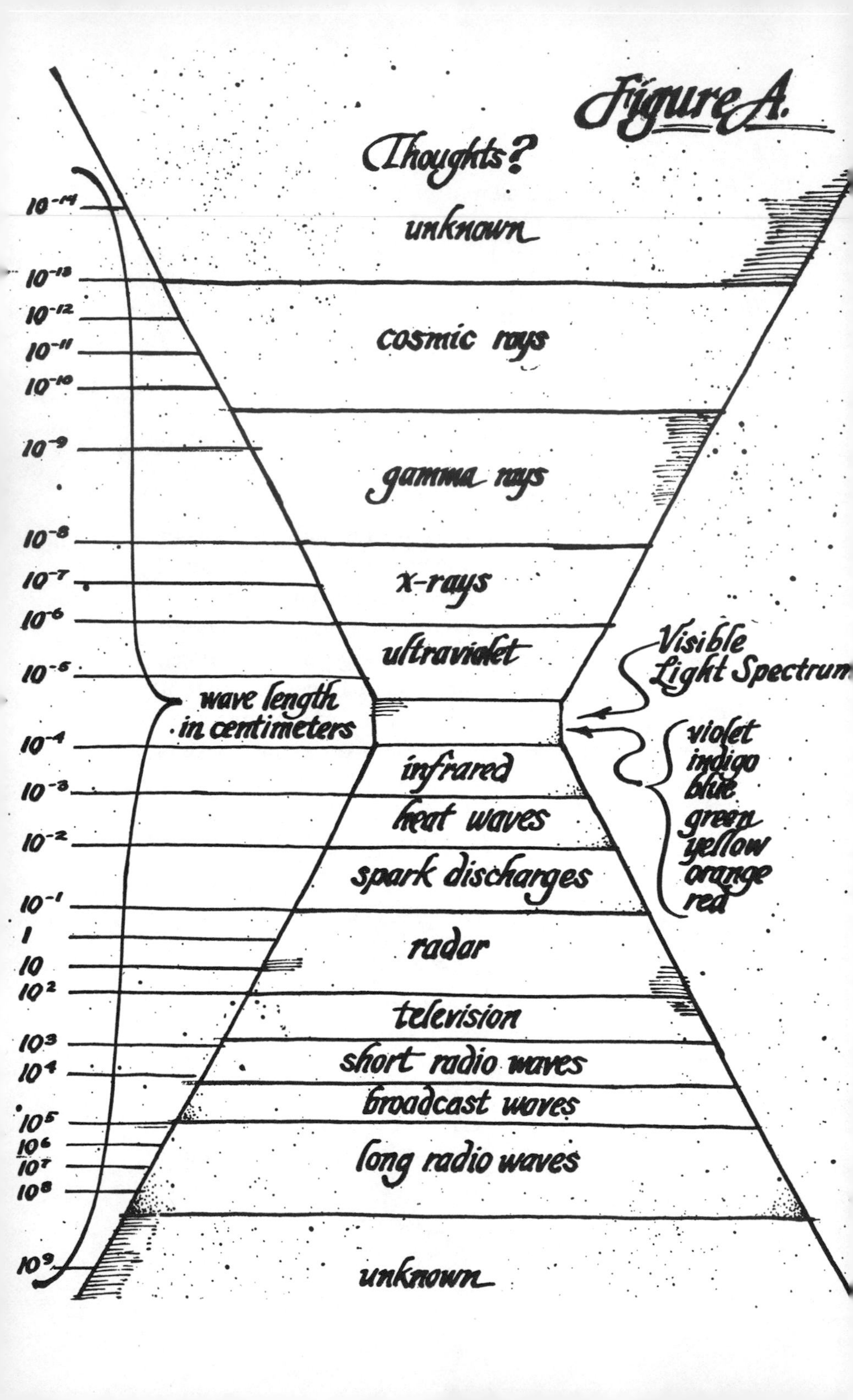

Figure A.
Thoughts?
unknown
cosmic rays
gamma rays
x-rays
ultraviolet
infrared
heat waves
spark discharges
radar
television
short radio waves
broadcast waves
long radio waves
unknown
wave length in centimeters
Visible Light Spectrum
violet
indigo
blue
green
yellow
orange
red
10^{-14}
10^{-13}
10^{-12}
10^{-11}
10^{-10}
10^{-9}
10^{-8}
10^{-7}
10^{-6}
10^{-5}
10^{-4}
10^{-3}
10^{-2}
10^{-1}
1
10
10^{2}
10^{3}
10^{4}
10^{5}
10^{6}
10^{7}
10^{8}
10^{9}

The electromagnetic spectrum in Figure A indicates wave lengths by the denary system: ie., 10^3 centimeters equals $10 \times 10 \times 10 = 1000$; and 10^{-3} equals $1/10 \times 1/10 \times 1/10 = 1/1000$

Basic physics reveals that the only difference between radio waves, radar, visible light, x-rays, and cosmic rays lies in their rate of vibration. Only a minute portion of the vast spectrum of vibration is visible to the human eye, other vibrations, however, do obviously exist.

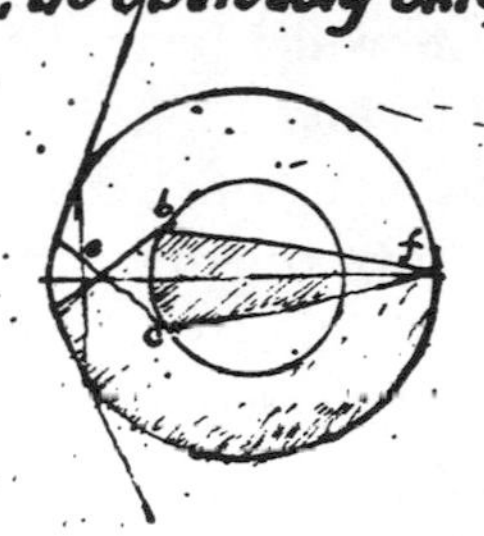

Figure B.

Thought? Sight Hear Smell Taste Touch Matter?

increase in vibration speed ← five physical senses → decrease in vibration speed

igure B. indicates the spectrum of the five physical senses— om touch (a slow rate of vibration), through sight (a higher rate vibration).

more. Like the propellor of an airplane, we know it's there but we can't see it.

At this point the pen is moving so fast you can't see it, but no one would deny that it is, in fact, still a ballpoint pen. Now we turn the speed of the machine up and the pen moves even faster—so fast, in fact, that at this point physics tells us that we would begin to hear it. First we would hear a low pitched vibration, then one higher and higher until it is beyond the range of the human ear. It's still a pen, right? Let's move it faster; turn the dials on the machine up and really get that baby moving. Now you can't see it, and nothing on the planet can hear it—but it is still a pen. Everything that the pen is composed of is still there. (At some point some of the atoms will get quite hot, hook up with other atoms, and the pen may change its composition a bit, but all of the pen is still there.)

As the pen glows with a red heat you move it faster and you see it go through orange, yellow, green, blue, purple, violet, ultraviolet, x-rays, gamma rays, and beyond. It has gone beyond the human visual spectrum. Move it even faster, faster than trillions of cycles per second, and you have thought. Thoughts are ballpoint pens? We've heard that thoughts are things now for several thousand years, are we first begining to realize why? Thought and matter — eletromagnetic particles. In a sense I guess you could say that the pen is nothing but slowed down light. Everything in the universe is composed of the same thing vibrating at different rates, causing different densities. Thought is the highest form of matter; matter is the lowest form of thought.

Everything is in continuous change, constantly moving and flowing. The energy in the universe can neither be created nor destroyed; energy just is. If I were to take a paper envelope and put a lighted match to it, most people would say that the envelope is gone, disappeared. It is true that the physical form of the envelope would no longer be perceptible to the human eye, but all that was an envelope, all the particles that composed that envelope are still here, only in a different form. Energy can neither be created or destroyed. Energy is.

Another example would be to set a bowl of water outside. Eventually, through the evaporation process, the water disappears. Is it gone for good? No, the bowl will fill up again during a rain. This is the natural cycling process of nature. Just because we can't see the water doesn't mean it isn't there; the human eye cannot detect it. Scientific instruments can measure the humidity, and we do know it is there even though it is not visible.

A wise old sage once said, "To believe in the things seen is no belief at all. To believe in the unseen is a triumph, a blessing." These thoughts are reinforced by Jane Roberts and her Seth books. "As living cells have a structure, react to stimuli and organize according to their own classification, so do thoughts. Thoughts thrive on association. They magnetically attract others like themselves, and like some strange microscopic animals they repel their 'enemies,' or other thoughts that are threatening to their own survival." Seth.

Infinite Distributor

Your brain is an instrument through which the mind works and your body is directed by your brain. You are, in fact, a distributor of life energy. The very word man, as derived from the Sanskrit root, means distributor or measurer.

We know from experiments done on human body polarity that the body does have a current which flows from left to right, and with sensitive techniques such as kirlian photography, biofeedback, and the like it has been determined that the energy flow is definitely affected by thought or state of mind.

Enormous feats of strength have been displayed by individuals who, when the need arose, directed their energy into a specific task. Back in 1960, a 123 pound woman, Mrs. Maxwell Rodgers of Tampa, Florida, lifted a station wagon weighing 3600 pounds off her son; he had become pinned underneath the vehicle. Many similar instances have occured with young and old people alike. Where does the energy come from? It's always there, but for some people it seems a super human task would be just getting out of bed in the morning.

It is possible to squash or surpress this life energy through failure, depressed or non-survival type thinking, but this doesn't limit the energy, it only limits our use of the energy. Energy is neither created or destroyed — energy is!

Now, we as human beings are able not only to distribute energy, but to direct it to the area of our choice. This is a real challenge, for the energy will flow through us whether we direct it or not. Using this energy effectively in your life is what this book is all about. Only you can do the "squashing" in your life, because you are in control of the direction of the flow of your thoughts. Affirm "I am now willing to release the thoughts and things that clutter my life," and realize that you are literally plugged into life. The public service company won't cut you off for lack of payment, and it is up to you to decide what you are going to do with this power.

Your Magnetic Personality

Have you ever wondered just exactly why some people seem to attract people, laughter, money, and other successes while others repel these gains? Have you ever wondered why some people seem to always be in tip-top shape, fit as a fiddle and loving life, while others just can't seem to get over their cold or the flu, and catch everything that comes along?

How about the people who never seem to have much, if any, mechanical difficulty with their car—in contrast to the other guy who's always got his in the garage? You see and hear people with a real flair for music, acting or dance, while others have nothing. Some people are always tired; others have boundless energy. Some people are inventive, imaginative, and creative; others are stagnant.

The answer to these differences is vibration. Whatever you hold in thought, like a magnet, it is drawn to you. When you think anything, you magnetize or draw to you those particles of energy which are harmonious, resonant or in tune with the state of vibration of your thought. This blend of energy provides the necessary ingredients to produce a thought form, or thought prototype; a seed has been planted in the "garden of your mind." as Emerson called it.

A Day With The Farmer

When a farmer goes out to his field to plant corn, he prepares the soil, lays out the rows, and digs the holes for the planting of the corn. Then he drops in the seeds and covers them up. Each day he may go out to the field to see what's happening, watering the seeds a little and watching, then watering and watching some more. After a short time has passed he goes out to his field and finds it filled — with banana trees. Ridiculous? Right! He prepared the ground for corn, he planted corn, he watered each day with a calm assurance that he would get corn. He would not find his field filled with bananas.

Yet this is exactly what we do to ourselves all the time without thinking: We plant one thing and expect another. We vibrate in thought, drawing to us exactly the thought's harmonious counterpart, for better or worse ... vibration!

I often hear people say things like, "Boy, did I have a tough day; I'm beat." The mind says to the brain, "Okay, guys, he's choosing thoughts of beatness and we sure don't want to slough off on the job. Let's get to work and make this guy beat." So the brain sponsors all sorts of action, such as slower circulation, shallow breathing, a need to yawn, a feeling of heavy eyelids and drowsiness. If the victim didn't get to bed until 2 A.M. and he's got to get up at 6 A.M., he is probably lying there looking at the clock and thinking, "It's 2 A.M.! I'll be so tired tomorrow morning I'll never be able to put in a decent day's work. Even if I were to fall asleep right this very second, which I'm not going to do, I'd still only get four hours of sleep". Time passes and his mind worries about not getting any sleep ... until it's 3 A.M. "Oh my God, now I've only got *three* hours!"

The next morning, when his alarm goes off (the most unnatural way to wake up that I know of), he looks at the clock and thinks, "Boy, am I tired! I didn't get any sleep last night." Actually, he got three hours—and his thinking is not going to acknowledge that. When he gets to the office and someone says good morning, he says, "Don't talk to me. Am I tired! I didn't get any sleep last night." He goes through that day affirming his tiredness; everything he does he filters through his "tired" strainer.

Most people don't even have to go through all that to get tired. They just look at their "tired machine" — the watch on their wrist — and automatically regulate their feelings to the time. "It's 11 A.M., 30 minutes past break time. Boy, am I tired." "Boy, am I glad that day's over." The fatigue may not be due to hard work at all. Often it comes from boredom—and when he gets home he'll find out that he's not tired at all.

For example, our friend may get home, throw his coat on the couch, and say to his wife, "Gosh, honey, I'm really beat." Then she informs him that they are going dancing in an hour. "*Dancing,*" he says. "Not on your life." "But, Sweetie, we *planned* this!" However reluctantly at first, he begins to think about the fun they will have, and says, "I can be ready in about thirty minutes, honey." Now he has planted a seed of energy and aliveness, and it too will manifest, as did the tired seed. Vibration!

Let's Play Catch Up

How many times have you said to someone, "I've got to catch up on my sleep"? Did you know that you can't do that? For some reason, if we miss a few hours of sleep one night we think we have to "catch up." But catching up is not possible, and, I might add, not necessary — the body is a perfectly functioning mechanism, and it decides not only how much sleep you need but what kind of sleep.

There is good sleep and bad sleep. Studies indicate that the good sleep is your dreaming sleep, or REM (rapid eye movement) sleep, of which you only get about 1½ to 2 hours in a regular 7½ to 8 hour night of sleep. Dr. William Dement, of Stanford University, found, along with other sleep researchers, that 1½ to 2 hours of good sleep a night seems to be all one needs. This was determined in a sleep laboratory by having the subjects go to sleep while wired to an electroencephalograph. When the patterns on the graph indicated that the subject was dreaming, the researchers would wake him up. After a few days of not being allowed to sleep during the dreaming state, but allowed to sleep only during periods of non-dream (which amounted to, in many cases, as much as 8 to 10 hours a night), the subjects became irritable and psychotic.

The reverse was tried, allowing the subjects to sleep during the dreaming periods and awakening them during the non-dreaming times. Actual sleep time per night was only 1½ to 2 hours, but the astonishing results were that the subjects didn't seem to need any more sleep. They were able to function fine on that amount of sleep, and go about their daily activities of work or study.

Interesting lay experiments have been carried out by people wanting to establish records for sleeplessness. A disc jockey from New York stayed awake for eight days; a seventeen-year-old boy from San Diego stayed awake for eleven days; another disc jockey from Maine stayed awake for eleven days and four hours. Why? To find out just how the human body and mind work. After all those days of staying awake the longest period needed for the body to "catch up" was only thirteen hours of sleep. Some only slept eight hours after the ordeal. Those sleeping periods, however, were almost totally in a dream state. The new record for staying awake is now up to eighteen days, seventeen hours. Obviously many people are making themselves more tired than necessary through misinformation. We never run out of energy, we just misuse it by choosing thoughts of tiredness.

The body takes what type of sleep it needs without our direction or worry. It is much like eating an apple. You don't know what your body needs from that apple, but your body does, and it takes what it needs and gets rid of the rest. It doesn't store everything. The same is true with sleep. The body knows what it needs and without your help it gets just that. If you miss a few hours of sleep, the only important sleep you missed was dreaming-sleep—the 1½ to 2 hours we seem to need. The body knows that it missed out on some of this "good" sleep and it will compensate for this either sometime during the same night or possibly the next night by providing you with more dreaming time. You don't even have to worry about it.

6
Filing Your Flight Plan

I conceive that when a man deliberates whether he shall do a thing or not do it, he does nothing else but consider whether it be better for himself to do it or not to do it.

—Thomas Hobbes:
Questions Concerning Liberty, 1656

Have you ever planned for a vacation? Of course you have, even if it's only been a one-day vacation. I know many people and I'm sure you do, too, that plan extensively where they're going, what sights they'll see, what highway to take, how far they'll travel each day. They will first figure out how much time they have to spend and then proportion their trip accordingly so as to return on a certain day. It's amazing, but quite true, that most people spend more time planning their vacation than they do their own lives.

These people not only don't set long range goals; they very often don't even know what's going to happen during the day. I asked a man in one of my seminars what was going to be the first thing he'd do when he arrived at the office on Monday and he said, "I'm not sure. I guess I'll wait until I see what really needs attention." I call that "crisis management." You hop on the first thing that flares up. This management style is not the least bit unusual; this is the way most people handle their day.

I am quite sure that the greater majority of people do not plan for failure in life, but the irony is by not planning, failure is often the result. It is known that most people, by the age of 65 years, are dependent upon someone else for their support. It all looks good while still young and even on into midlife but suddenly the realization comes that planning for the "later years" had not been done. It has been said so often that if you don't know where you're going, you'll have a hard time getting there.

Even if a goal is set, self-discipline, or lack of it, is often one of the major factors involved in failure. Thus a habit is formed of letting ourselves down or we fall into the familiar trap of getting sidetracked. In either case it is a blow to the self-esteem and each instance helps to build a failure pattern.

Come Fly with Me

A friend of mine is a captain for United Airlines. He has always loved flying and to this day, even though he's been on the job for 23 years, he seems to find each day as thrilling as the first. He's the kind of guy who takes nothing for granted and leaves nothing to chance.

I asked him one day what he did in preparation for a flight. He said, "First you've got to know where you're going." After that he said he'd check the weather forecasts for the destination and alternates, find out how much fuel is necessary, consider the possible baggage and passenger list for the weight and balance computations, check the winds aloft and estimate the time en route. He knows before he starts just where he is going, how much fuel he'll need and approximately how long it will take him. Then he takes off with that specific destination in mind.

Let's say that my friend, while flying from Denver to San Francisco, gets a craving for lobster and decides to put down in Salt Lake City for an hour or two while he takes the crew to dinner. Just a short time of course; he's not really planning on staying long. After dinner they all board the plane again and continue on their way. Now the Grand Canyon is just south of Salt Lake City, not really directly on the way to San Francisco, but with the Boeing 727 it's really just a short hop, and besides it would be so beautiful at sunset this time of year...so off they go.

Not very realistic, is it? But that's just the way most people function in life. Even if goals are set, more often than not they are never reached because people tend to get sidetracked. I believe that you can be, do or have anything you want in life as long as you have a direction and purpose in mind and you don't get sidetracked. The clock of life is wound but once. Too often we reach a point where we say it's too late to get a fresh start or that we can't go any further. I believe that the person who makes such a statement has no place in mind to go and consequently is going nowhere fast.

Tuck a bit of this information away if you will. In rest homes for the aged the death rate drops drastically just before holidays, weddings, anniversaries, birthdays and other special events. Why? It seems that many of the people set a goal to live just one more anniversary, birthday or Christmas. Having something to live for makes life more valuable.

My father's recent actions provide a good example. He had lived in Milwaukee for most of his seventy-some years, and he noticed that some of his spark had dwindled. Being on top of things, as he usually is, he decided some changes were in order. He phoned me just a few months back and said he'd spent 77 winters in the cold and snow, and he was ready for a warmer climate. It didn't take him long to spring into action; before I knew it his house was sold and he'd bought a new one down in Fort Myers, Florida. He moved down there in the dead of winter and he loves it. It's never too late to change!

Very often people will say that they are not interested in material things; they would prefer to seek happiness, love or peace of mind as a goal. Eleanor Roosevelt said that happiness and peace of mind were never a goal; happiness and peace of mind are the *results* of having goals. Nevertheless, these two comprise the state in which most human beings desire to live. We must realize that before we can gain an understanding of how we can work toward and obtain something we cannot see, we must understand how to obtain those things we can see. Regardless of the goal you set for yourself, remember that something for nothing equals nothing. Whatever it is that you want from life, be it physical, mental or spiritual, you must first decide what you are prepared to give up for it.

Give It Up

Let's say your goal is to make more money. You could possibly consider giving up an extra hour at home so you could make that extra sales call. You could give up an extra hour of sleep in the morning (if you did this every morning for one year you would have nine extra forty hour weeks). You could give up drinks and a large meal at noon so you are awake and alert in the afternoon. You might consider giving up some of your leisure time to upgrade your education in your present job or another. You might want to increase or decrease your social life, depending on the contacts made and the contributions made toward attaining your goal. But rest assured, for every goal you set you must be prepared to shift gears a little to achieve it; if you did nothing at all different you would remain right where you are, and of course the goal would never be reached.

Priorities

Some people claim they'd really like to do a particular thing, but that they don't seem to have the time, or for some reason something else gets in the way. Obviously you must have a desire to do something or it won't get done. If you truly desire to do something, absolutely nothing can stand in your way. If you really desire to reach your goal, then you will find the time necessary to do what you need to do. Remember, we all have the same amount of time.

There is a perfect guideline as to whether you desire what you say you do: are you moving toward your goal? Again I hear "but...but...but..." "I have too much else to do," or "I don't have time." Would you find time in your busiest day to have lunch with Burt Reynolds or Farrah Fawcett? Priorities—*set them!* Go for them; meet them!

Tomorrow Las Vegas

Suppose your best friend called this morning and said, "Hey, I've just been given two free tickets to Las Vegas. The plane leaves tomorrow and I want you to go with me." Assuming this is an attractive proposal, you'd probably begin planning immediately just how you could take advantage of this trip. Out would come the pencil and paper. What to do first? What can be delayed until you return in a few days? What to pack? Time schedules, phone calls, dog sitters, plant waterers, bills pending. For the next 24 hours you'd be a whiz of efficiency as you set priorities and arranged the details of living so you could say yes to the trip. This is exactly what I did in this very situation. Just after my friend called with those tickets I put everything into shape and we left the next morning. We did the town, ate great food, took in the Rich Little show at the MGM Grand Hotel, and had a great time. The reason for this is simple. A goal was set, a commitment was made, an organized plan of action was developed, and I jumped in and got started. Is it possible for *you* to get that excited and organized about your present goal right now? Other people can stop you temporarily; only you can do it permanently!

The Big Put-Off

"Never put off till tomorrow what you can do the day after." Some people actually follow this rule, and procrastinate their lives away. It takes zero energy to do nothing. When you've set your goal, the first day has to come in the new plan of action. If you do nothing, ask yourself what happened. If you hear yourself saying, "Nothing, oh, I should have started yesterday. I promise, I'll do it tomorrow," ask yourself what's wrong with *now?!* What you didn't do doesn't exist. It was never done. You see, you either do something or you don't. You can never "should have done" something yesterday. Should provides many people with an acceptable excuse for not doing something. If I know I was a bad guy for not doing it, I'll say, "I should have done it," and then everything will be okay. Often we feel defeated when we accept as true what others say we "should have" done. The classic bumper sticker was the one I saw the other day: "Have you been should on today?"

There's Still Hope

If you should have done it and didn't, the next step usually is hope. Hope is somewhere between wish and doubt, neither of which is any good. "Try" is another one: "I'll really try." You may as well say right off that you're not going to do it and at least be honest with yourself.

I am well aware that there can be blocks to reaching your goal and it would be unnatural if there weren't. The only time any of us will be without problems is when we're six feet under. Strangely, it's not the problem that causes the problem, but how we react to the problem.

The Ditch

I was walking in the downtown section of Denver one sunny, hot, and dry afternoon on my way to a business appointment. Most people seemed to be moving slowly because of the heat. I spotted a maintenance crew working on a section of the street, and one man was in the trench shovelling the dirt out, and each shovelfull was coming out faster than the one before it. I thought to myself, "That poor man, someone must have been buckling down on him to make him work so hard, especially in this heat."

I made my way over to the trench and peered down at the man, who was covered with perspiration and dirt, and said, "Looks like you could use a break."

"Break." he said. "I've already lost ten pounds this month and I'm working on ten more. Don't bother me buddy." Obviously I did not see the situation in the same light as he did. I've come to the conclusion that there is neither a good or bad sitation; it's all in the way you look at it.

When you bite into an apple your taste determines whether it is sweet or sour, but consider this—if you've been eating honey for an hour the apple will taste sour. On the other hand, if you've been eating lemons the apple will taste sweet. Life, like the apple, can taste sweet or sour and your effectiveness or ineffectiveness has much to do with your attitude.

499 Pounds or Less

Recently I was in New York to attend a program in which Arnold Schwarzenegger spoke. He related an incident which impressed me. Arnold Schwarzenegger, the body building expert, told of the effort put forth by weight lifters the world over to lift 500 pounds. Whether in competition or training it seemed a physical impossiblility to lift 500 pounds. 490, yes! 498, yes! 499, yes! But not 500 pounds. Then in 1970 during the Weight Lifters World Championship event in Columbus, Ohio, Alexis asked for 499 pounds, the current record. He jerked the weight over his head. As always in a championship event, the weights were weighed again upon completion. When placing them upon the scale it was found that the first weighing was in error; they didn't weigh 499 pounds, but, in fact, 501½ pounds. A new world record was established.

Within six months, six other weight lifters lifted over 500 pounds. Why? Better training? Changes in food? Drugs? No! Nothing changed other than that people now saw it was possible, and because they were mentally convinced, the physical body responded. I do believe psychological barriers were broken, providing a free path for others to follow—that once 500 pounds was lifted it was accepted as a possibility and others subsequently followed in their path.

You Can Have Anything You Want

Wow! What a powerful statement. "Anything I want, he says. I don't believe it." And that's where most people stop. They never go deeper into the meaning of it.

Yes, I believe you can have anything you want. A statement as bold as that certainly takes some explaining, for as you know, most people don't believe that's possible. In fact, it's hard enough to get people to believe that they can and ought to go after the things that are rightfully theirs.

Habit plays a big role in this plan of goal setting. William James, often called the father of American psychology, said, "To change becomes increasingly difficult and requires purposeful effort to counteract a system already set-up." Notice that, although he acknowledged the difficulties of change, he did not say change is impossible.

It seems that the older a person becomes, or the longer he remains in a particular system, the more dfficult a switch from that situation becomes. An old saying makes the point quaintly: "Refrain tonight, and that shall lend a kind of easiness to the next abstinence, and the next more easy, for use can almost change the stamp of nature."

Put off, put off, and put off, and all of a sudden it becomes so easy not to do it. In other words, the habit of *not* doing is formed, rather than the habit of doing, which successful people form.

Fear, the Monster Motivator

For a person to be able to make an intelligent decision as to whether something is right for him, it is first necessary that he become aware of the reasons on which a decision is based.

Fear of failure, fear of heights, fear of love, fear of rejection, sickness, poverty, and even fear of success are often the basis of not doing things. The list of fears could be endless, for there are many, but it seems that one of the biggest fears of all mankind, and by the way, one of the major reasons why so many people fail to set goals, is the fear of change.

Fear of giving up the old shoe, of moving ahead, and pressing onward and upward is unbelievably strong in some people. Even if the new situation we're looking toward is for our advancement and betterment, the fear of change is often too overpowering. You see, we become complacent in our everyday ways. We become settled and comfortable with what we know; we fall into the well-known rut. For a lot of people, a rut is a grave with the ends kicked out.

Maslow said, "We either move forward into growth or back into safety." He adds, "The unrest, the unhappiness and the unease in the world today is caused by people living far below their capacity, and if we plan on being anything less than we're capable of being, we'll probably be unhappy all the days of our lives." Often we become satisfied with where we are in life because it's easy. We know the job and change would require effort. Change may mean putting yourself in a situation where you suddenly feel stupid again, and the fear of what others will think becomes acute. Many people refuse to take chances in life, because they fear the criticism which may follow if they fail. The fear of criticism in such cases is stronger than the desire for success.

Many people also refuse to set high goals for themselves, or even neglect selecting a career, because they fear the criticism of relatives and friends who may tell them they're being unrealistic — that they'll never be able to accomplish their stated aims.

Far too often we have put other people's opinions of what we ought to do above our own opinions of what we think we ought to do. To be your own person in our society today requires determination, for everyday we are bombarded with ideas, slogans, music, and other people who are dedicated to making us all just like eveyone else. This effort is directed toward preserving or forming comfortable habits — which will resist change.

Mental Habitude

Certainly there are good habits and bad habits. The trick is to keep the good and restructure the bad. Habits are often helpful; for instance getting dressed in the morning could be a real bummer if you had to think about tying your shoes each time you got dressed. Many repeated behaviors become advantageous habits, because they allow the mind the opportunity to do other things.

The trouble begins when, through habit, we prejudge — and this means both people and situations. By prejudging we slip everything into a convenient slot and do not allow ourselves the opportunity for growth or freshness. It's hard enough to judge a person or situation correctly when you have all the facts: it's impossible to judge accurately beforehand.

With Ease or With Difficulty

Assuming that you have set a goal and you have made a commitment to move in that direction, how do you know for sure that the goal you have set is right for you? Just a few paragraphs ago I said you can have anything you want. You can — but you could get yourself into a lot of trouble if you bring things into your life that are really not right for you.

A major factor in reaching a goal is *desire.* If you are striving and fighting to assist someone else in reaching their goal, that's nice, that's kind-hearted, that's being a good person and we'll pat you on the back for it, but you must not lose sight of your own desires. Sometimes it is possible to get the desires of your loved ones mixed up with your own.

Isn't it great that we are all different and have our own desires? What a world it would be if we all wanted to be zookeepers, surgeons, or accountants. Life speaks to us through desire; whenever you desire something, that's life knocking at your door saying, "I'd like to do this through you. How about it?" The rest is up to you. If you don't pay attention to your desires and fulfill them, no one else can. If you don't sing your own song, it won't be sung at all.

The first step in deciding whether the goal you set is right for you is quite simple. Ask yourself if you *really* desire to achieve that goal. The next step is to look at the ease with which your goal is accomplished. If everything falls your way, if there are no blocks and no barriers, if everything moves smoothly, the path to your goal is easy. If a thing is easy it will also probably be fun to do. The ease of reaching the goal is a determining factor in whether or not it is right for you. If accomplishing the goal flows easily from the desire, it is probably right.

Persistence is important, of course, and if accomplishing a goal doesn't come easy, you can work at it until you reach it. If you are proceeding with great difficulty, experiencing great strain, resistance, and frustration, you may be pursuing something which is not right for you. To clear this up we have to go back to the beginning, the desire. If you truly desire something, you will also desire the way it is to be accomplished. You will not have to fight your way to the top. The way to the goal may still be diffficult, but your outlook will be considerably different. My own rule of thumb for measuring the rightness of goals which prove difficult to reach is to ask myself, "Is it fun?" If it is, I'm probably on the right track; if it's not I fall back and take another look at it. Possibly the goal is right but your method of accomplishment is wrong.

Order

I used to think goal-setting was contrary to the working of the law of mind. I thought that living in the present moment meant just that. Don't worry about the future; don't set goals; don't plan. If you lived your present moment that's the best you could do, and by setting goals you were trying to control the future. However, history reveals that where there is no vision, the people will perish. It seems that you must have a picture in your mind to bring your desires into being. So what's the answer?

The Naturalness of Nature

Order seems to be nature's first law. It is natural to think in pictures, and thinking in pictures must be nature's orderly way of thinking. To think of your goal, to put it in your mind in terms of a present possibility and hold that in your mind, all your actions and thoughts between the time you set the goal and the time you reach the goal must be naturally goal-directed and necessary to bring your goal into being.

If you are living each moment out of your true desire, enjoying it, loving it, and above all, are excited about it, you will be pursuing your goal in a natural, orderly way. Setting a goal and living in the present moment are not contrary to one another at all. They are a perfect blend, a natural blend. Enjoyment is the key. Your goal may be right for you whether it is easy or difficult to reach, but the way must be enjoyable! Success will be the result, for the press of nature in all that she does is for expansion and fuller expression. Advancement into all things is nature's great purpose.

Nature does not seek lack, deterioration, disease, depression, or failure. The laws of life are all life-directed, and life means growth. Growth in turn requires change, and as we move through life using enjoyment as our vehicle, we will see our goal successfully unfold before our eyes.

7
Sales Aloft

If you want to succeed you should
strike out on new paths rather than travel
the worn paths of accepted success.
—John D. Rockefeller

The art of selling is used by each of us everyday of our lives, because we all deal with people. No one can be part of the game of life without being involved in the game of selling. Whether we are selling someone on an idea, selling a product or selling to groups or individuals, it's all the same—we seek that magical blending of two or more minds into a unified idea. At the very base of the selling process, however, lies the factor which determines whether your ideas will be accepted or rejected. This base building block upon which all future sales is built is just this; selling yourself on you!

The SIP Principle

Self-Image equals Performance: SI=P. When you sip a glass of iced tea you sip slowly and gently. That is just what you do when you begin selling yourself on you. Let's examine why it is so important to feel good about yourself in sales before we get into the method of accomplishment.

To begin with, no one can make you either a good or bad salesperson. Only you can do that for yourself. However, outside influences often enter in when we don't want them to. Ideas may be thrust upon us by others: "You haven't made a sale in weeks." ("You're terrible.") Do we accept it? Sometimes we do, and what happens as a result? We may go around moping, feeling dejected and generally down in the dumps. Now, I ask you this: at that point would you feel like going into the corporate executive's office and asking for the sale? Probably not. On the other hand, if you've just received a plaque for outstanding salesmanship and everyone is congratulating you, how do you feel? Wouldn't this be a better time to tackle that corporate executive? Of course it would! The only difference is that your self-image is in better shape, that's all. You don't know anymore about selling, you just feel better about yourself.

In this example the salesman would be allowing something outside of him to affect his behavior. Even though this often happens, it is entirely possible to choose not to let it. That's why it is so important to learn to pat yourself on the back and provide yourself with the inner self-assurance that is essential in today's business world.

Can you see from the above example that whenever you raise your own opinion of yourself you raise the level of your performance? Conversely, when you lower your opinion of yourself, you lower the level of your performance. Outside influences are not the cause of your effectiveness or ineffectiveness. They are simply a condition of the world around you. As a good friend of mine says, "You've got to feel super fine."

Once you do, you can take on the world, for there is no stopping a person who knows his own self-worth.

If a thing can be done at all, there are usually several ways of doing it, and it seems logical that some must be better than others. Selling is not a hit or miss proposition; there are rules that can be followed to assure a successful end.

In a sense selling is like pounding a nail into a piece of wood. Tap gently a few times to ensure the placement of the nail and then lift the hammer high over the nail and pound firmly, thrusting the nail into the wood. After a few strokes the desired result is achieved.

A good carpenter never "muscles" the nail into the wood. He takes advantage of the tool and lets the hammer's weight and lever action drive the nail in for him, thereby enabling him to work long hours without tiring. He guides his tools to do his job, and the more he understands the principles by which they work the easier his job becomes.

Understanding the principles by which sales are made is extremely important to the salesman. There are tools which can be used which can never meet with failure, and, just like carpentry, selling is a skill and can be learned.

SIP It

The tool we will be using will be mental. Unlike the physical nail being forced into the wood, you can never force anything on the mental plane. Instead of force, we use awareness — awareness of how things work. Sip at it gently and surely, for awareness begets more awareness and little by little you can see the growing process building right before your eyes.

The foundation on which this is based begins with the understanding that everything — animal, vegetable, or mineral — reflects patterns of energy. Realizing this, let's say that you desire to sell something to someone. You know now that the other person is composed of the same life energy as you and therefore has all the qualities available to that life energy. You see, everything contains everything, the highest of highs and the lowest of lows. All the qualities you desire are there, such as understanding, decisiveness, openmindedness, receptivity, etc.These qualities may not be readily apparent in the present moment, but they are there and the proper tool must be used to bring them out.

Oh, for Wayne!

When we see a streak of stubbornness in someone, often we'll say to ourselves,"That guy couldn't see a good deal if it hit him in the face." A word of warning here; condemnation of any kind is a definite roadblock to sales. The most effective tool is to acknowledge what you want from the individual, not what you don't want.

When I was a kid living in Milwaukee, my best friend was Wayne. We spent many good years together, playing kick the can, hide and seek, spooking the neighbors on occasion, but generally keeping out of trouble as much as is possible for two young boys.

We both lived on North 26th Street, and I remember strolling down the street toward Wayne's house in the mornings, walking through the yard and standing at his back door calling for Wayne—"Oh, for Wayne!" Why I didn't use the doorbell I'm not sure; it just wasn't the thing to do in those days. You stood outside and shouted for all to hear, "Oh, for Wayne!" Guess who came out? Not his mother, not his father, not the neighbors, not his sister, Audrey (although I often wished she would). No, Wayne came out. Why? I was calling for him, that's why!

The simple message is that you get what you call for. In sales, if you are calling stubbornness or other uncooperative attitudes out of your customer that's what you'll get. On the other hand, if you begin to mentally call for receptivity to the product, openmindedness to a new way, willingness to listen objectively, you'll see a different side to that person. You can stand there and not say a word and watch the person melt into a pleasant attitude right before your very eyes.

This is in no way forcing an opinion from the customer. Rather, it is bringing to the surface his best qualities, thereby allowing him to better make up his mind after all the facts are presented. We know that the life energy flows through everything. It powers your customer as well as it powers you. The life energy in the buyer and in the seller, in the dog catcher and the dog, in the bookstore clerk and the bookbuyer, in the dress designer and the manufacturer, in the mother and the child, in the automobile and the television and ultimately in your customer are all one in the same life energy in an individualized expression. Can you see the unity this presents?

Now a paradox is presented with this power or force, for to bring this power to a binding oneness and unity two poles are required, blending and balancing. The poles in sales are selling and buying. One power conceives of the idea to sell through you. But to complete the cycle it must also have a buyer. One cannot be complete without the other, just as there is no temperature without hot and cold, no color without dark and light, no up without a down. Whenever you experience a desire to sell, you need only know that the buyer is already made available; the only thing you need to do is recognize the sale as already existing. It is a law of sales that a desire to sell must be balanced by a desire to buy. One cannot be without the other, and they will draw together like iron filings to a magnet to complete the one desire.

The belief that separate powers exist for each, for both customer and salesman, is the great pitfall of many. The realization of the unity of oneness that binds the two poles together is the key, one power acting as both parties.

In short, don't call out faults in your client; call out the qualities that you desire expressed. Realize the underlying unity between both parties and acknowledge that the completed sale awaits, with eager longing, to be revealed to you. Fall into the flow of successful selling by using all available tools, so that, like the carpenter, you can work without tiring.

PART THREE

intuition

8
The Hidden Compass

A vague feeling, perhaps discomfort, a flash,
an idea, too often diagnosed by logic as
indigestion or fantasy. Small wonder any of us
hear the intuitive voice within that produced
light bulbs, atomic energy and Polaroid cameras.
Successful listening is the key to our minds.

—Jim Williams:
Denver, Colorado,
September, 1978

Many people look doubtful when the word psychic is mentioned. Why? Because it's a misused, misunderstood, and loaded word. The sixth sense, hunch, insight, intuition, extrasensory perception—these are some of the synonyms for the word psychic. Even in today's world of deep mental exploration the psychic nature of the individual is sometimes shoved into the closet of the occult. Again, lack of understanding is the major reason for this. Intuition means to be tutored from within, that's all.

So to begin with, let's get this thing called intuition out of the closet and into the living room where we can look at it in broad daylight. All creatures that roam the planet, including humans, are intuitive.

In the lower forms of life we call this faculty instinct. A finer degree of knowing is inherent in humans, present in each of us. How many of us, for instance, have had a hunch about who was on the other end of the phone before we even answered it? Or sitting next to someone, you begin talking at the same time as the other person—about the same thing. Occasionally you may have a feeling that you should or shouldn't do something, and only later realize why.

Insights into people are another excellent example of the presence of this power. Any employer knows that much of the hiring process consists of a basic "I feel right about you" sensation. Most employees can tell whether a raise is coming even before the final word is spoken. A management consultant to businesses which collectively represent sales of about $60 billion yearly, states that in dealing with these pragmatic, successful enterprises, his counsel is based 80 percent on intuition, 10 percent on logic, and 10 percent on experience. Furthermore, he states that all the succesful people he's associated with rely heavily on intuition for decision-making. Extensive research into the psychic nature of the individual has been carried out at Duke and Stanford Universities, and examples abound to show how, and in what forms, this power exists.

All of us are intuitive, and all of us express this power to some degree. I'm convinced that we all could, if we chose, develop this power to a higher and finer level. My interest is in exploring the principle itself and developing it into a reliable measuring stick for use in everyday life.

Attention! Right or Left Face?

You may have heard of the right brain, left brain concept. The human brain is divided in half, into a right and a left brain. The right side of the brain controls the left side of the body and the left side of the brain controls the right side of the body. However, the significance of the division is also that the two halves are not merely mirror images, but perform different functions as well.

In this society we focus mainly on the left brain. The left brain functions deal with reason, intellect, logic, decision-making — facts and functions in a linear sense. On the other hand, the right brain is intuitive, creative, artistic — it functions in a non-linear way. Both functions are necessary, but we don't often get a blending of the functions in our society.

As a left-brain society, we are technologically advanced and analytical; we calculate and investigate just about everything. In itself that's not bad, but by itself, it's very unbalanced.

Eastern cultures are much the opposite. They tend to be not so technically advanced; they are considered to be more spiritual and intuitive. In reality everyone has all the qualities of the brain; it is only the emphasized development of one side or the other which causes the imbalance.

One of the major reasons why people in Western cultures don't consider themselves intuitive or psychic is that they put little energy or concentration in that area. It is known that what you focus on will tend to manifest itself. In addition, men often reject their intuitive side as unmasculine — intuition is often referred to as "women's intuition," and in most cases this is not acceptable to men, who are taught that they should be factual, objective, and unemotional.

One of the places that the imbalance is striking is in the business world as well as in government. Many women working in these areas end up in the same trap as the men, attacking everything with logic and reasoning power. To some degree this is necessary, but the intuitive softness, creativity, understanding and feeling nature — "women's intuition" — is badly needed in the business world today. It seems to be difficult to pull these qualities out of the business mind, be it male or female.

Tapping intuition has been the undertaking of many researchers, scientists and theologians, and the ability to use this area of the mind has remained a mystery until quite recently. The true mystics, however — the people who are in touch with themselves and the universe —have always been able to display a sense of insight beyond the so-called norm. But the average person has been afraid and unsure of exploring the psychic realm.

We now seem to be breaking into a new age, one that I believe will prove to be one of the most exciting in history. Hopefully it will become commonplace for every person—men as well as women—to discover the power within himself, and in this process many of the long held secrets of the universe will be rediscovered. I say *re-discovered* because it seems that much which was previously known has been lost. By investing in the true nature of man, as opposed to just the physical trinkets he makes, the great enigma of life can then at least be questioned.

Child's Play

Children are natural psychics. Children do things by feeling; they don't have to have a reason. They do things for fun; they are free spirits, not governed by rules and regulations but by what they feel at the moment. They live in the present moment and don't struggle to live in the past or strain to live in the future, and when children are not encumbered by a lot of do's and don'ts, they seem to have a basic knowledge of what's good or bad for them.

When left completely alone to select his food a child will select a perfectly balanced diet, not by the day of course but by the month. Children are capable of intuitive eating, intuitive playing, intuitive exercise, and yes, even intuitive sleeping; a child knows when he's tired and he will sleep. Naps and a long night's sleep may fit into the adult's schedule, but they may not be what the child needs. Perhaps we should consider adjusting our sleeping patterns to coincide with our children's.

Children have the ability to construct vivid mental pictures, but they are usually told to stop daydreaming. Many even go so far as to develop a friendship with an imaginary person—until they are told that that is nonsense. A child's mind is a wonderful thing, and when today's adult is trying to become more aware and intuitive, in many ways he is trying to become a child again.

Reliable Results

A person who displays considerable psychic ability has several qualities. One of the most prominent is that he has confidence in himself and his ability. It has been found that people who are sure of their ability in this area are more accurate than those who are unsure of their ability. Another characteristic is that a psychic will often use first impressions as being correct. When the mind dwells on a subject it tends to begin the reasoning process and this is definitely what you don't want during the psychic process. Quite often you'll find that the psychic will be in a relaxed state of mind and body and more often than not will be in the alpha brain wave state.

Performance on demand does not often bring out a psychic's true powers. In fact, undue pressure usually inhibits the functioning of the psychic considerably.

I do not believe that psychics are different from auto mechanics. Both have a talent; each has a particular interest. If you are interested in auto mechanics you will pursue that line of work; if you are interested in the psychic realm you will pursue that course. Neither ought to be in awe of the other, for each could do as well in the other's field if he so *desired.*

What's Ahead?

Since the past is over and the future is not yet here, all we really ever have is the present moment. Our future is really made up of a collection of present moments. If you want to be a psychic and read into the future, all you really have to do is to look at where you are now, for most probably you will continue along the same lines unless otherwise acted upon by some outside force.

In the psychic realm there is no time or space; all is here and now. Let me explain. Your probable future is that you will finish reading this sentence. The improbable future is that you'll jump right up from where you are, leap on a white stallion and go charging down the main street.

Somewhere in between lie all sorts of possibilities. You're the boss. You can choose what to do at any time, and there isn't a psychic in the world with any credibility who could even begin to tell you differently. We are not predestined; we always have free choice in every matter.

The late Thomas Troward once said that the one Truth in the universe can be found in the Bible, a Pack of Cards and the Great Pyramid. I believe that it can also be found deep in the eyes of a loved one, in the white cumulus cloud, the falling aspen leaf, and even in this book — for some. Each person will communicate with Truth in a different way, and there are as many answers and ways of seeking as there are people on the planet. There are no right or wrong ways, just different ways.

If you are intuitive to a larger degree than others and you are able to help yourself and others with this information, something is obviously working for you along those lines. My advice to you is to keep doing what's working.

9
Your Intercom System

'Tis the uncreative mind that can spell a word but one way.
—Paul G. White, Jr.

It's nice to know how to spell, but the real purpose of writing is communication. Whether you spell city with a *c* or an *s*, it sounds the same.

I was sitting in a restaurant in Milwaukee one noon waiting for my lunch when another customer in the restaurant got up to pay the bill. She was an old woman, gray-haired, dressed in long sweeping pieces of cloth draped around her, and she had on a lot of make-up. She approached the waitress and said quite clearly, "To what is the extent of my indebtedness?" I thought, what's wrong with, "How much do I owe?" Then I thought again and said to myself, what's wrong with what she said? After all, the basic idea is communication; who cares how she says it as long as the person she's talking to understands?

Spelling and speaking correctly—why all the fuss? I do understand the need for a basic language, but there will always be variations in its basic structure. What's "right" in one part of the country may never have been heard of in other parts. Even the way words are pronounced varies considerably, depending on where you're raised—in the South, the Bronx, Milwaukee, or Italy. The whole basis of language is communication; the main objective is to transfer a meaning from one person to another. One effective tool we use is verbal communication. However, as Emerson put it, "What you are speaks so loudly, I can't hear what you're saying." This seems to indicate that we speak with more than just words; we speak in consciousness.

Instant Hate

Have you ever, upon being introduced to someone, developed an immediate like or dislike for them? Maybe they haven't even spoken yet, but you've already formed an opinion. It seems that we carry with us a presence of attitude that is perceptible to others as well as to ourselves. This has been explained by some as an energy discharge surrounding the body—a radiance, an emanation from each living individual. Often this presence is referred to as the "psychic" or "magnetic atmosphere," but the most common reference to this presence is the term "feeling," as in "I just had a feeling about him." Most persons are more or less aware of that subtle something about the personality of others, which can be sensed or felt in a clear, although unusual, way when the other persons are nearby, even though they may be out of the range of vision. Being outside of the ordinary range of the five senses, we are apt to feel that there is something queer, something uncanny about these feelings of projected personality. But every person, deep inside, knows these feelings to be real and admits their effect upon his impressions regarding the persons from whom they emanate. Even small children and animals perceive this influence and respond to it with liking or disliking.

Sundance

A friend of mine raised Sundance from a pup. Sundance was a Doberman Pinscher. When Freddie moved to another state to take a job which required him to do extensive traveling, he gave Sundance to a friend, knowing that his dog, who was very dear to him, would be cared for. A few years later he returned to his home town and went to visit his friend. He walked up to the house, opened the fence gate, and went into the yard—and Sundance came running up to him like a long-lost friend. It was good to see his dog after so long.

When his friend opened the door he asked Freddie how he got by the guard dog in the front yard, and pointed to the "Beware of Dog" sign on the gate. Freddie said, "Well, that's Sundance, and he remembers me."

"I'm sorry," the friend said, "Sundance was run over by a car two years ago. I got this guard dog because we have had trouble with prowlers."

On the way out, Freddie's friend offered to walk him to the gate. Freddie said, "No, I came in; I'll make it out." And make it out he did, but just barely. You see, now he knew that the dog wasn't Sundance but a guard dog, and the communication, even though unspoken, was different between him and the animal.

Certainly communication with others is important. But what are you communicating to others? Just what is it you are saying? Take a good look at yourself, and ask this: "Would you like to be your best friend?" If the answer is yes, you're probably in pretty good shape, but if you'd like to clean up a few areas before you take on you as a friend maybe this will be of some help.

Word Choices

The impression you communicate to others is the exact impression you have of yourself. So, what kind of person are you? What makes you tick?

Make a list of things you like and don't like, and on it put such things as sports, candy, people, men, women, cats, auto mechanics, reading, movies, wind, snow, swimming, and so on. In other words. make a real list. Put down about 25 items in all, just of various things that come to your mind. Then go back and in a column next to the word indicate whether you like that thing.

Now make a column next to your list—this one an "I'm good at this" column or an "I'm not good at this" column. Again write down 25 things, and then go back and indicate whether you're good at each one. Don't be selective; just write whatever comes to your mind. Your list may contain such things as carpentry, running, eating, loving, dating, reading, gardening, drawing, and so on.

When you've completed the lists and commented on each item, realize this: you are forever continuing to form your likes and dislikes, what you're good at and what you're bad at, by the way you continually think and talk about it. I often hear people say things like, "Well, that's me. I never was any good at math." Then they go on to affirm, "I even have trouble adding." The statement, "I am," draws a check on the universe and the proceeds go to you. What are you saying to yourself?

Look at your list again. Notice that there are some things on it that you indicated you weren't good at but you'd really like to be good at. Here's where word choices come in. If you would like to change your experience in an area — say, for instance, your list indicates you're no good in cooking, but you'd really like to be a better cook — the way you choose to talk about cooking will affect the end results of your cooking. How is that possible? Well, "I am a bad cook" is a quite powerful statement. With the word choice technique, we flush out the old saying and replace it with a fresh one: "I am a good cook." That's ridiculous, you say. Well, perhaps it isn't.

First of all, if you like what's happening in your life obviously there is no reason to change it. However, if there are areas that you'd like to change and do better in then word choices are important to restructure your thinking about that thing. I agree that it would be ridiculous if in fact you are already a good conversationalist to repeat on a continual basis "I am a good conversationalist." But if you consider yourself an awkward conversationalist and you stumble in front of people groping for words, never seeming to have the proper response to anything, you may want to change. What you say about it creates a response as to how you feel about it. And as you know by now, your thoughts determine your feelings. For so long you have chosen to select thoughts of being a poor conversationalist that you have become firmly tied to that belief. You have drummed it into your head and the results are disastrous. "I am a poor conversationalist." Remember, the universe always says *yes*. It is impartial and will provide you with anything you choose to be.

The mind, acting on a feeling, sponsors an action that is normal, natural, and automatic. You don't even have to think about it. You just are. If you keep affirming that you are a poor conversationalist, all the patterns for being a poor conversationalist fall right into place; you don't even have to work for it. Why? Because the mind is acting on the feeling of a poor conversationalist.

The word choice technique works like this. Imagine yourself standing on a cliff and shouting into the canyon, "I am a good conversationalist." You get an echo back that says, "No, you're not. You talk too fast and you don't dress well." The echo reveals part of the problem. Perhaps you *do* need to slow your speech down a little; perhaps you also want to check into what it is about your dress that makes you not want to talk to people.

Continue with the word choice technique each day and soon you'll root out all the reasons that were in your way to success in that area. Now, again, continue with the statement of "I am a good conversationalist" and observe any slight change for the better. Constantly watch for any possible sign of improvement. Take every opportunity to experiment with your new beliefs by gradually entering into conversations — slowly at first, and also selectively.

Make it a point to interject a statement or two, especially if you are used to not saying anything. Observe the response of the people and again, re-excite the mind with, "I am a good conversationalist."

It's so easy to label yourself as being either one way or another and then use that as justification for remaining the same and never even having to change. Whenever you make a word choice of "I am," in essence you are saying "That's it. Don't even try to change me."

If the "I am" is for your benefit, obviously you will want to keep it, but if the "I am" is keeping you from experiencing more of the good in life or holding you back in any way from expressing the fullest possible you, then change the word choice and flush the old one down the drain.

Concerned Listening

Verbal communication sometimes takes more thought behind it than one might think, or than one might desire to give it. Concerned listening is often a helpful process, for through it someone may be able to bring to the surface a deep-rooted problem of another or, shall we say, expose the underlying message.

Concerned listening can be used when you feel that the other person, be it adult, child, boss or employee, is saying something that doesn't quite ring true—something that you feel could be covering the real reason for his or her unpleasantness or unhappiness

Usually we listen to someone with only one ear, not really giving the words our full attention. With concerned listening you fully listen to the person in the deepest sense, and then you put into words what you hear that person feeling. You don't parrot back to them exactly what they just said; you repeat it in terms of what you just "heard" from their feelings.

An example might be when your child comes home from school and says, "I hate school." Your first impulse might be to respond on the order of, "You have to go to school; if you don't get a good education you'll never amount to anything. The least you could do is be grateful for the opportunity to get an education." That's called *unconcerned* listening!

Much of *unconcerned* listening amounts to advice giving. Oh, how we all love to give advice. It's so easy to do. A little more time is needed when it comes to concerned listening, because we want to get to the root of the reason for hating school—if that, in fact, is the problem. So your response could follow these lines: "You really feel bad when you're at school, do you?" Child says, "Well, some of my classes are okay, and I like recess."

Parent: I'm glad that you like some of your classes. I always thought recess was fun, too, when I went to school.

Child: Well, I really don't dislike any one class in particular. It's just that school takes up so much time.

Parent: You find the long hours hard on you?

And on and on until finally the real reason is revealed. The child says, "Well, since you started that night job, I never see much of you and school gives me even less time at home."

This answer would never have been found by the advice-giving parent. A caring, understanding and concerned parent will use concerned listening to develop insight instead of oversight. Married couples will find concerned listening a most effective tool to root out petty differences as well as other possible impending barriers to communication. Just asking your mate to be a concerned listener to you is a good beginning to opening the channels for a more effective relationship.

Concerned listening does not mean that you judge what the other person is saying. You don't try to interpret it or analyze it or put the other person down for feeling any particular way. You just listen and repeat back what you heard the other person feeling.

You will also find a whole new response being expressed to you by the other person. Everyone, to some degree, likes to talk about himself, and that's exactly what he's doing. You're not imposing any of yourself into the conversation; it's just the other person. Concerned listening can be quite effective.

Say It If It's So

Do you express to others what you feel? Most people don't. They more often couch their feelings in words that they think will be more in line with what someone else would like to hear. Words are also used to make another person feel guilty, such as, "You always spoil the party," or "If it weren't for you I'd be able to go." This is not what I mean by "Say it if it's so."

Letting someone know your real feelings often puts the conversation on an entirely different level, enabling real communication to take place. Instead of criticizing the other person's shortcomings (*e.g.*, you did this, you made a mess, you don't care anymore—

you, you, you), it can be helpful to state how *you* feel about the situation. For example, "When you don't stand up straight, I feel like walking fifteen feet behind you." A brief statement of the behavior and then how you feel about it is often all that is necessary to correct a situation you are uncomfortable with. You don't have to dig for solutions. Let the other person provide those, be it child or adult. By letting him know how you feel you've done all you can without interfering, and by letting him choose to find the solution you have given him an opportunity to build self-esteem.

Cumulus Granite

While we are in a continual stage of development, occasionally there seem to be blocks thrown in our way, making personal growth a little harder than at other times. This is readily demonstrated by the basic learning curve of an individual. When we approach a new subject the learning is absorbed quite rapidly. We take in all the new bits and pieces of information and store them for future use. After the initial phase the person reaches a point where the learning seems to slow down, and almost stops at times. At these plateaus many people become discouraged and call it quits. "What happened?" they wonder. "Everything was going so great. I was learning so much and then *pow!* Nothing."

This is a point at which the "seeming" block occurs. It is really a false block to growth; it only seems like a stumbling block or slowing down process. Actually this is the time for absorption. It is time to assimilate what you've already learned. It's like eating food—you don't keep eating constantly. You take time to digest what you have eaten, and then later on you eat some more.

There are, however, real blocks to growth, and until those are moved out of the way they'll keep you stopped dead in your tracks. When I lived on the West Coast I did a lot of flying. There was a weather condition that was a definite block to flying; we called it cumulus granite. That's a mountain. When a mountain gets between you and your destination you do one of two things. You either go around or over it, or you plow right into it. Now the latter is not recommended; it becomes quite hard to continue the flight when your nose is sticking into the side of a mountain.

There are also blocks to personal growth that can be just as disastrous as the mountain is to flying, but in our society these are covered up so well that it takes a little looking to discover them.

Dis-cover. To take away the cover so you can see. And we do become blind to so much in our world by falling into the trap of following others and not thinking for ourselves. We see a thousand people going one way, and just a few going the other. We say all those people can't be wrong—but they often are. It seems that the higher in consciousness and awareness you are the more alone you become.

A major block to personal growth occurs when we place someone else's opinion of us above our own opinion of ourselves. We see people falling into that trap all the time. Someone can say, "I think you are a terrible artist." You may accept that as being true, or you may consider it as just being one person's opinion. Someone might call you stupid. But someone else's assertion that you are stupid doesn't make you stupid. It would be as ridiculous to think that as to think that if someone called you an elm tree you should try to be an elm tree. And yet we continually let others govern how we feel and how we respond to situations in life. We become the dumpee in life, letting others dump on us their own non-survival behavior patterns.

A Matter of Choice

What others think of you is only valid to the degree that you choose to accept their opinions. We have total choice in everything that we do. We choose our clothes, our food, how much sleep we want, the type of work we do. We choose our mate, pets, friends, where we'll live. Why be unhappy, then, with where you are in life. If it is all a total choice, then we can also accept or reject what other people say about us. After all, each of us is entitled to an opinion just by virtue of the fact that we exist.

If You Don't Like It, Then What?

I was in Los Angeles not long ago talking with a cab driver and I asked him how he liked the city. He said it was terrible. There were too many people, it was noisy, the traffic was bad, and on and on. I asked him where he'd like to live, and he said Denver. I got to Denver and asked someone how he liked living there and he said it was terrible, there was no place to go, it was too quiet, and so on. I asked him where he would like to live and he said, "Chicago." I got to Chicago not long after, doing one of my programs, and I asked someone how he liked his city and he said it was terrible. There was smog, it was windy, humid, too hot, or too cold. Yet each of these men had chosen to be there. What's the point in not liking where you are? It's where you've chosen to be. If you don't like it, why not move? It seems that most people would rather complain about something than take action to change it. What they don't realize is that if *they* don't change it, it's not going to get changed.

Don't Touch Me!

Communication by voice is probably the most accepted method in our society, but even though it is effective it is limited. Touching is used very little in our culture compared to other countries. The impression somehow has been drummed into us that to touch is to attack.

Certainly there are acceptable ways to touch. There is the handshake; among intimates there is a kiss or a hug. But even those are only "permitted" on certain occasions. It's all in what you're used to. In the U.S. we're not very "touchy" people. In Mexico, South America and other Latin countries, touching is a way of life. I've traveled in Mexico and the attitudes toward physical closeness seem to be much more relaxed. I remember in particular standing in a nearly empty train station. A long bench was empty except for one person sitting toward the end near a small table. Another person came over and sat down within touching distance of the other, and it didn't seem to disturb either of them. They both continued to wait for the train and to read.

A study was done at the University of Denver regarding invasion of personal "territories."

Psychology students were asked to go to the library and locate a table where only one person was sitting, select some books to read and then, at this large table, sit down right next to the lone person. In most cases, the person under study would become nervous and would soon move to another table or leave the room. Why? Most people are not used to closeness. Our culture does not promote it, and body contact is often completely impermissible.

I'm not saying our way is right or wrong; I am saying it is different. In many other countries a common greeting for both men and women would be a kiss. In some countries it is not uncommon to see a man walking hand in hand down the street with another man. In the U.S. the two would be considered gay. In Iran, if when conversing with another you are not close enough to feel the other's breath on your face, you are too far away.

I believe that a little more closeness could be the threshold of a little more understanding with our fellow man. It seems to me that touching is quite important, and often we hold ourselves back rather than express our feelings. We have four senses contained within the head: sight, hearing, taste, and smell. The sense of touch encompasses our entire body. To a large degree the sense of touch is a safety device, protecting us from injury, but it is also one of the most important senses that we have. Not a part of our body is without it.

The Monkey

Laboratory experiments with monkeys proved that touching is necessary for life. Immediately after birth the monkeys were isolated, and from that point on had no physical contact with either human or monkey. The young monkeys soon became despondent, depressed, and sickly, and ultimately they died.

Touching can be, and most often is, a healing process. Touching is one of the warmest ways to express love. Caring for another, be it animal, plant, loved one, friend, family, or fellow human being, can be expressed best by touching, even without words. Touching is probably the most effective way to communicate in this physical world.

I have always been given to touching. It's not uncommon for me to talk to someone else with my hand on his or her shoulder. I've always been quite free with hugs. I do, however, take into consideration the other person's feelings, for I realize that invasion of privacy can be most uncomfortable and, to some, even embarrassing. Uneasiness can always be sensed and should be respected. Just because someone is not comfortable with touching does not mean that they are not caring. A need *not* to touch can be just as great as a need to touch. It all lies in our belief patterns, environment, upbringing, and so on. I think, though, that the need to touch is the more natural way to go.

Glass Bubble

That uncomfortable presence can be felt almost wherever you go. Of course, it doesn't have to be uncomfortable. We just make it that way. You can notice it especially in the elevator and you can sense a glass bubble surrounding each person. This is their "space." They are entitled to it. You notice it even more in larger cities where crowds are greater. You notice it on the subway, on the bus, standing in line for a movie, even sitting in the dentist's office with one or two other people.

Now, if you are the type that notices there is a glass bubble around people most of the time, try this. Take the glass bubble off yourself and relax. You'll see an enormous change in others. Release that uptight feeling, and mentally send out energy of warmth. Practice radiating a presence of communication as opposed to non-communication and see what happens. You may be in for some surprises.

A friend of mine who lives in New York is an architect. He used to take the subway to work, but after hearing the glass bubble idea he decided to walk — mainly to get the exercise and encounter more people. It took him 40 minutes to walk to work. He started at his front door by taking some deep breaths, standing up straight and then beginning a brisk walk. When he made eye contact with someone, he smiled. Walking to work became an experience for him. It lightened his day, and by carrying an attitude of aliveness on his way he encountered alive people, making his walk much more pleasant. Bob carried the same attitude into the office, and he said that before long he actually began to enjoy his days there. In fact, he said that it wasn't really like working. It was more like going to visit his friends for eight hours and then walking home. Now that kind of job sounds great!

This poem by James Allen seems to sum it up all very nicely.

You tell on yourself by the friends you seek,
By the very manner in which you speak,
By the way you employ your leisure time,
By the use you make of dollar and dime,
You tell what you are by the things you wear,

By the spirit in which your burdens bear,
By the kind of things at which you laugh,
By records you play on the phonograph.
You tell what you are by the way you walk,
By the things of which you delight to talk
By the manner in which you bear defeat.
By so simple a thing as how you eat.
By the books you choose from the/
well filled shelf;
In these ways and more, you tell on yourself;
So there's really no particle of sense
In an effort to keep up false pretense.

Let's look at the first line a little more closely. "You tell on yourself by the friends you seek." Let's say that you are attending a funeral and you don't know the person who is lying in the casket, but you are curious as to what kind of person he or she was. All you need do is to get acquainted with the friends who have come to mourn, and that will give you a pretty good picture. You see, like attracts like. If you are a corporate executive you probably don't hang around with the bums in the Bowery. If you are on the honor roll at school you don't chum around with the dropouts. If you are a healthy person you probably have no patience with hypochondriacs.

This doesn't mean that some people are good and some people are not. People just are—good, bad, or whatever. It means that we tend to choose friends and be around people that are like ourselves.

Consider for a moment how you use your leisure time. TV is the main leisure "activity" for many people. I've heard it said that TV is a vacuum tube you sit in front of to create a head to match. A recent study determined that between the ages of 2 and 65 the average North American will watch an average of nine years of television. Amazing! What do you do with *your* leisure time?

"You tell what you are by the things you wear." Clothes don't necessarily determine what a person is like inside, but they often give a pretty good indication. Generally it can be said that you dress the way you feel. The whole key, then, would be to get yourself feeling good, then looking good, then smelling good, as Reverend Ike says.

The way you walk can also communicate your attitude to others. It's not uncommon to see someone on the street who goes first one way, then the other, back and forth. These people don't seem to know which direction they are going, and what's worse, they don't seem to care. On the other hand there are people who are obviously going somewhere. There is no time wasted; they get straight to the point. They have an idea in mind and they are going after it. You can sometimes tell just by the way someone walks.

What you laugh at and talk about is also an indication of what goes on in your mind. It's easy to gossip, isn't it? Keep in mind that there is no right or wrong, there just is. Also keep in mind that if you ever want to draw a different type of person into your life you must first begin to think differently. If you don't like the chatter that goes on in your life, remember that others only echo what you whisper in secret.

What you eat also displays your mental attitude, as described in the discussion of yin/yang concepts. The books you read do the same thing. What's on your shelf right now? What are you choosing to put into your mind? What movies do you choose to see? You are molding your life by the impressions you choose for your subconscious.

Your world is but a reflection of your thoughts. Choose them wisely!

10
Love and Marriage: A Flight Forever?

Who travels for love finds
a thousand miles
not longer than one.
—Japanese proverb

Marriage

A grand and glorious relationship, an unbeatable team, excitement beyond description, and then...*pow!* What happened? Gone forever? This is the story told by so many people in this up-dated, modern, do your own thing world. Distilled, watered down, dissipated love: What to do, where to go, how to handle it?

Is there such a thing as the "perfect marriage"? That question can only be answered by you, for no one can stand in your shoes and get behind your eyes to look out at the world. You are the one who knows whether you are happy.

I've gone through some rough times, but I wouldn't trade them for anything, for they have been extremely good lessons. I've made mistakes — more than some, fewer than others. But to look back on a mistake with condemnation is totally self-destructive.

There is an old saying which is appropriate here: "There is no sin but a mistake and there is no punishment but a consequence." Mistakes are guide points through life. They are opportunities for growth, so that we do more than just count birthdays as we advance through life.

Mid-Course Correction

A space flight to the moon begins with blast-off into orbit around the earth. All telemetry being in order, the space vehicle departs earth's orbit to proceed on a planned trajectory to the moon. Mid-way between the earth and moon a correction point is laid into the flight plan to provide an opportunity to correct the course if necessary. This gives everyone involved in the operation the chance to evaluate the position of the spacecraft and determine the number of degrees of correction needed for the craft to continue on its scheduled flight path.

Assuming that a mid-course correction is needed, and it usually is, do you suppose the astronauts berate and condemn themselves for having to correct their course? I have never heard of an astronaut saying something like, "Oh, my God, I made a mistake in my flight path. I was off course and I had to correct myself. What a fool I was!" A little ridiculous, you say? Most people do exactly that in correcting themselves on their path through life. Mistake means sin, and the root word for sin means to miss the mark. That's all it means. Someplace, someone goofed. Now it's time to correct it.

I do believe that the "perfect marriage" is possible; I also know that it is not often expressed, for to be happy with your mate you must first be happy with yourself. Or, as Gisele MacKenzie once put it, "Ladies, I've got news for you. Your husbands are not here to make you happy; you make yourself happy. Take your happiness to him and he'll add to it." I liked that one because it seemed to let me off the hook. In reality, of course, it works both ways.

Yet so many people are frantically searching about for happiness, looking for people that will make them happy and understand them. You sit in your living room looking at your wife or husband and you realize that even after ten or twenty years of marriage they don't even know you. Resign yourself to the fact; no one will ever know you. Part of life is an aloneness that will go on for each of us because no one can get behind our eyeballs and look through our eyes and view life from our perspective. It's just not possible. In addition, as true as it is that no one will ever know you as closely as yourself, so is it true that you will never know anyone else, for they too live in the same space within themselves.

The question is asked time and again, "What is the way to happiness?" But it's the wrong question, because happiness is the way. You either have it or you don't; you can never go anywhere to find it, for then it will always be someplace else. Happiness is the mode of transportation that we use to move through life.

Brush Fires

Have you ever put out a brush fire? Little fires flare up all around you. You throw a shovelfull of dirt on the biggest flare; then one on the other side of you gets bigger so that you turn to that one. You beat that one for awhile and suddenly another one springs up behind you.

Some people handle their personal lives the same way, carrying crisis management into the home. A financial situation gets out of hand so all your attention goes in that direction. Then quite unexpectedly the car falls to pieces. Then the job begins to suffer a little and demand more of your time. It's either the kids, a friend in need, an illness—you name it, something flares up and that's where your attention goes.

All this time your attention is driven away from home, or to put it more specifically, driven away from your marriage. That fire is also building. Too much is often taken for granted. Just because five, ten or fifteen years ago you had what seemed to be the world's best marriage doesn't necessarily mean that the same circumstances exist now. Certainly there is caring, certainly there is understanding, but just as every other situation in one's life needs attention, so does the marriage.

Living Together or Dying Together?

A flower that is living is growing, a flower that is not growing is dying. Life means growth. If you're not growing you're dying. Growth means change; to consider growing old together and never changing would be unrealistic. Nothing grows at the same rate in nature, and yet so many people expect to get married and have the exact same relationship all through life. This is as unrealistic as expecting that children will never grow up.

Just as a business needs constant attention to ensure proper direction in growth, so does a marriage partnership. Marriage patterns are similar to business patterns, but of course more personal. When you begin the business you lay out a plan for growth, exploring all the profitable and enjoyable ways to complete the task at hand. You try new advertising campaigns and promotional items, and in general you try to provide your clients with the best possible product by continual upgrading, all the time keeping in mind that if this is done your business will zoom to success.

Many people look at marriage as the final step. They see it as a point of completion rather than a point of beginning. The dating is over, the catch has been secured, and from here on in it will be smooth sailing. Their attention turns toward home furnishings, and in many cases children soon become a prime interest The man and woman are thrust into different roles and naturally experience different daily environmental patterns.

Consider the growth pattern of two children, both born on the same day in the same city to the same family. Ten years later you will find them to be quite diverse in their likes and dislikes. After twenty years, possibly, the difference may be even more striking, and yet their origin was essentially the same. A married couple's origin is also essentially the same, a point of unity when two minds blend, and sharing, caring and loving are of prime importance. Growth together begins. However, remember that growth means change, and no two things ever grow at the same rate. No two flowers, no two trees, no two dogs, and certainly no two people do or even can grow at the same rate. Allowances and adjustments are necessary ingredients in cultivating a continued, balanced, harmonious relationship. Exposure to new ideas and individually experiencing new situations creates a different attitudinal environment. To quote Oliver Wendell Holmes, "The mind, stretched to a new idea, can never return to its original dimension."

You can never go back to the old, for you will always carry with you your new experiences. Even if you did return to a former city, an old friend, a husband or wife, it could never be the same as it was before. Why? Change has taken place. Each has grown to some degree in some direction and you will remold the situation in accordance with the thoughts you hold in your own head. In short, you will experience what you decide to think about.

Love

I believe that love is possible for anyone and everyone at any time. Love is an art and, like any other masterful work such as music or poetry, it can be created. But like the arts, it requires focus, dedication, and discipline. I do not mean that you can go out and acquire love just for the asking, but I do mean that it is possible to apply oneself in a direction that would make it possible for love to come to you.

But what is love? Love gets so mixed up with inadequate feelings, insecurities, fears, inferiorities, sex, and money that it is a hodgepodge of confusion. Not very many people really know what love is. And yet it is something everyone is searching for. Hardly any other activity or enterprise is entered into with such tremendous hopes and expectations, and yet which fails so regularly, as love.

Our whole educational system presents the "facts" of life—mathematics, history, and so on—to provide a body of knowledge so one can become "educated." Yet one element—the main ingredient of life — seems to be sadly lacking: the love factor! Without it we lose. It is the thing all people crave and need to exist, yet it is so elusive that most people become confused and disheartened.

The problem is that many people don't know where to locate love. Furthermore, often the qualities of love are not understood. Flying to Los Angeles one afternoon I sat next to a very well-dressed, attractive, middle-aged lady. We began talking about the day and the wonders of flying, and then almost without warning she burst into a recital of her complete life story. Often a person will tell a perfect stranger things they wouldn't tell their own family.

Her husband had left her about three months before and she was returning from a vacation in the West Indies. She was traveling alone. She began to lay out all the reasons why she thought her husband shouldn't have left her: she had raised two beautiful children, made sure they were always well dressed and on time to school; she had maintained their home in an extremely tidy manner, practically dust-free. She was always ready to go anywhere he wanted to go; she was an excellent cook and had prepared some of the finest meals for her family, and she had entertained his friends, his co-workers and his boss. "All of these things," she said, "and he left me!"

The woman was searching for something and she didn't know how to find it. She didn't even know where to look. Everything she mentioned that morning was something her husband could have bought: a cook, a maid. a catered party. What is essential in love is something not found in the physical world. You can't see it or hear it; it isn't measured in the number of meals cooked or parties hosted. It is a feeling only to be experienced, not explained. We can theorize, practice, and concern ourselves with falling in love with someone, but before that is even possible a key ingredient must be added: a true, deep understanding love for yourself.

Those who find love are sometimes considered "lucky in love." It is a common phrase. However, I don't believe in luck. I've heard it said that luck is the alibi, used by reason to explain the success of intuition. The logical, intellectual and reasoning mind gathers facts and asks questions. The intuitive, creative and feeling mind non-linearally assimilates these facts and provides a gut-feeling answer. Intellect asks; intuition answers.

A person appears lucky when he follows his "feeling nature." Love is an emotional feeling and in itself cannot be intellectualized. However, a body of knowledge regarding love can be gathered, thereby providing the intuition with the necessary facts for it to act on.

Have you ever noticed how people are always going some place else to find love and happiness? "Where can I find love?" California?Chicago?Hawaii?a cruise?a party?a dance?So you go to the dance and you sit there and don't dance. Why? One woman said that there were no men there, just beasts and apes! She obviously was not entering into the spirit of things that evening. A certain degree of receptivity is necessary. A light-hearted, almost humorous attitude could be helpful instead of hunting and searching. It appears that those who fail in love most often are those who are seeking to be loved rather than to do the loving. People are looking for it "out there"; they say, "love me; I need somebody to love me." Placing your happiness in the outer world means that you also place your unhappiness out there. The following is a humorous example of the eternal search for love "out there."

A Maiden's Prayer (author unknown)

At sweet sixteen I first began to ask you/
dear Lord for a man.

At seventeen you will recall, I asked/
for someone strong and tall.

At Christmas when I reached eighteen,/
I fancied someone hard and lean.

Then at nineteen I was sure, I would/
fall for someone more mature.

At twenty I still thought I'd find,/
romance in someone with a mind.

I retrogressed at twenty-one, and found/
the college boys most fun.

My viewpoint changed at twenty-two,/
I longed for someone who'd be true.

I broke my heart at twenty-three,/
and asked for someone kind to me.

Then, longed, at blasé twenty-four,
for anyone who wouldn't bore.

Now, dear Lord, that I'm twenty-five,/
just send me someone who s alive.

The eternal search for love—is it to be found in money? "That'll do it; if I have money then I'll be loved." If that were true then everyone who was rich would have a wonderful loving relationship. But of course they don't! People alter their bodies surgically, tint their hair, buy expensive clothes, all in the name of love. Maybe it will help to find the "right person." Admittedly, the luxuries in life are nice to have, but you can never buy your way to love. You can buy sex, but to buy love is not possible!

Sex is a passionate interest in another person's body. Love is a passionate interest in another person's personality. Sex craves fulfillment and satisfaction. Love craves only fulfillment. You can force a person to sex; you can never force a person to love. Sex on the human level is always most satisfying when it comes as an expression of true love.

It is the epitome of sex expression when sex comes through love rather than just for physical gratification. Don't confuse love-making with love. There is a difference.

Why does anyone search for love? Ask yourself. It's not because you want to give something to someone; it is because you want something given to you. A man doesn't ask a woman to marry him to do her a favor. To feel worthy of this gift of love you must carry a high regard for yourself.

I've heard it put this way. We gather up all of our qualities, both good and bad, and put them in a package. We wrap it up, put a ribbon around it and then go out shopping for a mate, business partner or companion. We price our packages. Let's say that you have priced yours at $10.95. You will meet someone who is also out shopping with his or her package of qualities, and guess what? The price on that person's package will be $10.95! In order for you to attract someone with a higher priced package you must first raise the inner value you have set on yours.

A friend of mine had joined Alanon, a program designed for the families of alcoholics. She overheard two women talking and discovered that, between the two of them, they had had a total of five alcoholic husbands. They were wondering how that happened. Unless you elevate your awareness and self-image you will continue to attract the same type of person; only the clothes and the name will change. A situation will repeat itself until the message is realized. We either learn or repeat again, only to say, "Hey, I've been here before." When you raise your own opinion of yourself to the million dollar person that you really are, and price your package accordingly, the $10.95 person won't even be shopping at your store. Only then will you begin to get quality merchandise.

Again, like attracts like. I'm not talking about the physical world so much as I am about the world of personality. In the physical world many times we find that opposites attract. Although it is true that two physically attractive people will be drawn to each other, it is also true that physical attractivness is not the main ingredient to a loving relationship. In fact it doesn't have to much to do with love at all.

Have you ever met someone whose looks didn't impress you initially, but then after you had talked for awhile, they seemed to become more attractive? The reverse is also true. You can meet a most beautiful or handsome person and after talking for a while find that your interest has faded. It is the personality that we fall in love with, not the physical body. In the mental realm it *is* true that like attracts like.

The more things two people have in common, the greater the opportunity for a successful lasting relationship. Similarities such as religion, age, interests and tastes are key ingredients. If a woman who is cultured and interested in theater and the arts were to join forces with a man whose highest level of culture was Howdy Doody, how long do you suppose their relationship will last? Conversely, you may fall in love with someone who is far superior to you, but it is unlikely that they will fall in love with you. Most people are searching frantically for love and, like the coins at the bottom of the pool, you can't see it as long as you are splashing around looking for it. You must let the water become calm. Then it becomes quite clear where the coin is. The search for love is similar. Love could be sitting right on your doorstep and you might never see it. You can't buy or hunt it. It comes to you as a gift.

When I Gave Up

I wanted so much
for you to care
for you to want me
for you to think of me

I wanted so much
for you to think of me
pretty
intelligent
important

I wanted so much
for you to care
for you to want me
for you to think of me

When I gave up wanting
you cared

—Sheila O'Rourke Stearns
Long Beach, California

What Sheila is talking about here is release. One of the most powerful laws of attraction is to stop hanging on to that which you desire.

Have you ever tried to train a puppy to heel? When you have a training collar and leash on, the little pup will want to stop at every bush and tree to sniff and look around. All the while you are tugging and pulling at him to heel by your side as you walk. One of the best ways to get a puppy to follow you is to release his leash and walk briskly down the path, whistling and light-hearted—the little thing can't keep up fast enough. Moral: when you are happy and enjoying what you are doing, those around you want to be in on the action. Everyone loves fun, happiness and freedom. No one likes depression, tenseness and control.

A Diamond or a Zircon?

Love has many faces, and how to recognize it is the big question. To begin with, I believe that love tends to be geographical. If you live in California you will probably marry a Californian. Likewise, living in Wisconsin you'll probably marry a Wisconsinite. It seems that everyone has the capacity for love, so it must be a matter of cultivating love rather than searching for the "right person."

Assuming that you find someone that you are attracted to, how do you tell if the attraction is a diamond or a zircon? How do you tell if it's true love or infatuation?

I am not a believer in love at first sight. Infatuation at first sight, yes; but not love. Just like a little rose seed, love needs time to grow. It doesn't become a full-blossoming rose overnight. When I was a child I used to pick flower buds and force them open to find the flower. I was never successful. The flower was not there and neither was the fragrance, and in forcing it I killed it. A correct understanding of the law of growth is quite important here. To force, to speed up the process is to prevent the natural order of things. Nature's way is very orderly, and all that is necessary to bring forth a blossoming rose or a blossoming love will come naturally and in nature's own time. Success depends, then, on using the law of growth and not opposing it. The rose must be allowed to unfold naturally for it to become the beautiful thing that it is.

Like the rose, real love can never be found full-grown in a minute. Love, too, must unfold naturally through shared experiences, points of view, and hardships. To the uneducated eye a diamond and a zircon look alike. But to the trained eye there is a world of difference. Many are deceived by infatuation, thinking it is love. Infatuation fades with time; it wants what it wants when it wants it, and if it doesn't get it, infatuation is bored. Love, on the other hand, is more patient. To define what love is for each person would be difficult; each of us is not only a psychological, biological, neurological unit, but each of us also contains a special ingredient unlike anyone else in the world. This specialness is what seems to be the loving quality in a person, the thing that loves and is loved. It separates both you and the one you love from all the rest and never again can your rose look like any other rose, for yours is special.

With this in mind, I believe that *love is a true desire to see the one you love realize his or her highest potential.* It is an attitude, a point of unity, when two minds blend. It is the recognition of a strength, seemingly not present individually, created through caring and sharing of interests, viewpoints, and desires.

From this definition you can see why it is impossible to find love overnight. You cannot seek it or rush it, for you will kill it.

I once heard a story of two doctors operating on a man with heart trouble. One doctor was in training. The master surgeon said he would have to lift the heart to get to the problem area beneath it. In order to do this, an artery would have to be cut, the heart lifted, the problem area adjusted and the artery reconnected. All this would have to be accomplished in 45 seconds.

The doctor in training said, "But doctor, will we have enough time?" The surgeon's reply was, "We'll have plenty of time, just as long as we don't hurry!" Real love is never in a hurry; it is patient and forgiving and it grows with time.

Yes, love is available to all, with no limits attached, but we must create it. If you decide that you would like love in your life, you are the one who must first begin to do the loving. It must begin within yourself and move outward.

If you decide tonight to be a lover, it will take strength. It is the weak who are cruel; gentleness can only be expected from the strong. Patience, kindness, generosity, humility, courtesy, unselfishness, and sincerity are all qualities of love. A definite commitment is required to be a loving person, and once you do commit yourself you will begin to weave an enchantment beyond description. It's been said that "Happiness comes only when we push our hearts and brains to the farthest reaches of which we are capable. For the purpose of life is to matter, to count, to stand for something—to have it make some difference that we lived at all."

PART FOUR

prosperity

11
Winds of Wealth

Probably the greatest harm done
by vast wealth is the harm that we
of moderate means do ourselves
when we let the vices of envy and hatred
enter deep into our natures.

—Theodore Roosevelt:
speech in Providence, Rhode Island
August 23, 1902

The Breath of Life

"Beware of holding too much good in your hand." I couldn't believe my eyes when I first read that in Emerson's writings. What was he talking about? How is it possible to have *too much* good? It seemed as impossible as having too much health, too much happiness, love, money, help, or too many friends.

There I was sitting on the couch asking why, when I suddenly became aware that I was breathing. Breathing, of course, is automatic, but in this awareness there was a message for me. I learned long ago that when I don't know the answer to something and I am puzzled, it is useless to wallow in perplexity all day. Instead, I ask, "What do I need to know?"

My answer came through breathing; taking in air, letting it out. We hold it briefly, then exhale—a constant series of inhaling and exhaling. It's automatic; we do it and without even knowing we do it unless we consciously direct it. My big revelation was this: I had to let air out in order for more to take its place. "Beware of holding too much good in your hand." I had to let go to make room for more.

There is a rhythm present in the universe, a moving, pulsating presence of power, manifested in everything both physical and mental. It is predictable, measurable motion, to and fro, forward and backward, in and out. This natural rhythm is present in all things, seen and not seen.

If I were to hold my breath, I would consciously block the good, fresh, incoming air from entering my body. The breath of life would be stopped — not for long, because there is a back-up system. If I hold my breath long enough I will pass out, and then the system of nature takes over again. The universe is always for expansion and fuller expression. It is for life, not death, and we must get ourselves out of the way so that this natural power, ever present, can take over and give to us.

What Emerson was saying was that we can block our good by holding on too tightly to what we have. In order to get more we must be willing to let go of what we already have. It could also be stated this way: increase comes out of what we already have. This is why, for example, it is so important to begin sharing or putting back into circulation the money in your life—even if you don't have much.

Flow with the Current

There is no lack in the universe. The universe is totally abundant. There is an ever presence of ideas flowing through time and space, and it is up to us to take out of this flowing river an abundance of that which we want. The Mississippi River is a mighty body of water and it is made up of many other smaller rivers leading into it called tributaries; these tributaries assist the Mississippi in developing its strength.

There is only one Mississippi River. There is also only one universal law. However, it has many aspects, such as peace, joy, health, and abundance. Rightly used, these aspects can assist us in bringing our lives all the good that we desire.

The law only knows to create a positive result. Why, then, do we see so much which is negative? When we put in a negative, the law will create a "positive negative." Even a failure can be a successful failure.

The universe creates a mold and we fill it with our thoughts. You say something like, "I just can't seem to make ends meet." The universe says, "Okay. There you are. What else would you like?" You get just what you expect. The person who says "I can" and the one who says "I can't" are both right, in accordance with their beliefs. The universe always says yes! Standing in the flow of the universe and saying that there isn't enough good or money to go around is like standing in the middle of the Mississippi River and saying, "I can't have anymore water, all my water just went by." Even the word currency is taken from the word current. We just haven't known how to use it. We hoard it, hang on to it, steal it. Currency is meant to pass from hand to hand. This, being understood, can remove the dam and let the abundance begin to flow. It is imperative that the law of circulation be understood and practiced before the abundance that is rightfully yours can be experienced. It's as if you have a blank check. Not only are you not filling it out, but you are not cashing it.

The Force

I mentioned earlier that if I held my breath until I passed out, a natural back-up system would keep me going in spite of myself. The law of circulation always prevails. When we block it consciously through fear—fear of letting go of our money, fear of lack, fear of putting out, hoarding — a block is formed and that block must be removed, not just for our good but for universal expression. Remember, the universe is always for expansion and fuller expression. How can it function properly if there are blocks? So for its own self-preservation the universe must remove the blocks.

Some people never recognize the method by which this is accomplished. They end up putting their money back into circulation, but not at all the way you'd expect. When we hang on too tightly we end up recirculating through unexpected bills, like auto repairs, dental bills, shoes for kids, vet bills for sick animals, and so on. In general we are forced to circulate money whether we want to or not. Although the message is clear for all to hear, it has been said that the lips of wisdom are closed, except to the ears of the understanding.

Right now, look around in your life and see if there is any area in which you may be experiencing something similar to the blocks mentioned here. What is preventing you from keeping your money? Sickness? Roof repairs? Fender benders? Supposedly required donations? Gifts? Can you see any area at all where you feel you are forced into spending money? If so, ask the question that you don't know the answer to: what do I need to learn from this? In order to be open and receptive to an answer, I suggest that you review the chapter on intuition thoroughly. Asking is one thing, but hearing the answer is quite another. Remaining open and expectant are two of the keys.

Water Purification Plant

One of the greatest circulation systems ever designed is the natural recycling process of water. There is no new water on the planet. All there ever was is still here, recycled. It goes through various stages, of course; it evaporates, clouds become saturated, and rain falls. The water you bathe in tonight could be the water that filled the tub of Caesar. Think about that for awhile. The water you used to wash your car with last week may have contained the very same water that you drank as a baby. And the next bit of water you use may be mixed in the drink of the Inner Galactic Emperor of the twenty-fifth century.

There is no water shortage as far as the planet is concerned; there will always be the same amount of water — all there is. By storing it, redistributing it unnaturally, and containing it in various forms, we can disrupt the natural evaporation process and thereby create circumstances that seem to indicate a shortage, but in reality there is always enough.

The Human Valve

To open the valve of plenty you can use the same system as the water faucet. To get more water through the faucet you simply open the valve by turning the handle in the proper direction—you take away the barrier that is stopping the flow, and all the water you need is there for your use.

The valve of plenty in you can be turned on by turning your thoughts in the proper direction, thereby taking away the barriers that are stopping the flow. All the abundance you need is already there, ready for use. Choose your direction and then be done with it.

Once you've opened your channel and you have a clear consciousness of abundance, you'd like it to remain open, of course. Just like the hose attached to the faucet, it will water only those things in the direction it is pointed. If it's pointed toward rocks, it will water rocks and foster no growth at all. If it's pointed toward a garden it will provide the life-giving substance necessary for all kinds of vegetation. In the same way, you must pour out your abundance into areas of growth, feeding and replenishing such things as truth, success and all those God-directed endeavors of man. Water takes whatever form you pour it into, be it a bucket or a bottle. Likewise, thought is a penetrating fluid of the universe; pour it into the container marked "abundance." The universe always says *yes!*

Fill 'er Up

The drinking straw is a good tool for demonstrating how the very same principle by which a straw works can be used to get rich.

Most people think that by sucking on the end of a straw they are sucking the fluid up into their mouths. In reality, air pressure is forcing itself against the surface of the liquid in the glass, and the drinker by reducing the air pressure within the straw, creates a partial vacuum, thereby causing the liquid to be forced upward through the straw and into his mouth.

Vacuum pressure is also used in a lightbulb. The noise made when a bulb breaks is the air rushing in to fill it up. Higher air pressure outside the bulb is being forced into the area of vacuum. The same happens in a coffee can. When you break the seal the air rushes in. You can hear it forcing itself through even the smallest opening in the seal.

It's as if nature says, "Fill 'er up." Nature can't stand to have an empty space, and given any chance at all will force its way into the vacuum. Now, take the same lightbulb and imagine that it is not a vacuum inside, but instead it is filled with air of equal pressure to the air around it. Drill a small hole through the glass and nothing happens; no air rushes in. Why? The air is of equal pressure both inside and outside; nothing moves either in or out.

If you create a vacuum in your pocketbook, the forces of nature will tend toward filling that vacuum. Understanding this principle you can deliberately sow seeds and expect a good return. A good place to start might be your clothes closet. Take a good look at your clothes, and notice just how many articles you have not worn for months or possibly even years. Get rid of them. Create a vacuum in your closet, making room for new and better things. Give them away, perhaps to a charitable group that will redistribute them effectively. If some of the clothes are not worn out, someone else can enjoy them even more. If you haven't worn something in the last six to eight months, you won't be wearing it in the next six to eight months, either.

Don't be stingy, selfish, or greedy with your belongings. Greed is a scarcity concept, built on the idea that there is not enough to go around. It develops an expectation of lack. We live in a world of plenty, and in nature when a seed is sown we reap the harvest of many. Merely understanding this about money is, unfortunately, not good enough; we must use it. We must put the principle to the test, for only by this challenge can we see the results. Challenge by experimenting.

$15,000 Per Year

An insurance man whose territory included three counties in Wisconsin earned about $15,000 each year through his sales. At that time that was not a shabby amount. When an opportunity arose for him to enlarge his territory to about twice its present size, he jumped at the chance. He obviously would have more contacts because of the larger working area. After completing one year in his new expanded territory he totaled his sales and recorded $15,000 for his annual income. In spite of doubling his size of territory, his second year's yield was the same.

He was asked to train another man by bringing him into his area and letting him work with him. The new man began taking some of the clients and working various portions of the area. Even with the other man in his area, when our friend tallied up his year's earnings they came to about $15,000. It is quite evident that this man had a $15,000 a year consciousness.

He was quite puzzled as to why he was only making $15,000 a year when all the time his situation was changing. Finally he realized that he had a psychological block: his father made $15,000 per year. Our friend was taller than his father, smarter than his father, dressed better—in all areas of life he seemed to be doing better than his father, except one: money. The significance of this was that if he made more money than his father he'd put him out of a job. You see, making money was the only thing left for his father to counsel him on, so his subconscious decision was to keep his dad employed in the job of providing counsel to him.

He was asked to mentally give his dad a raise, to $25,000 a year. He thought of his dad as much more successful than ever before and being compensated for his efforts. He carried these ideas around with him during his working day and at the end of the year our friend's income came to $25,000. Isn't it strange the things we do to ourselves?

Jackpot

A good question might be, "How much can you accept?" This "acceptance" seems to be determined by our consciousness. Most of us know people who are just "rolling in it" and others who are "out on their ear" in the streets. Why the difference? If a man with virtually nothing is given a large sum of money, it won't be very long before he has nothing again. So many people think if they could just get their hands on a bundle their troubles would be over. They'd pay their bills, get totally out of debt, and have a chance for a new beginning. If you're in that pickle you'll probably find these statistics interesting.

Several years ago, at the beginning of the boom of TV quiz shows, the big money prizes were upwards of $50,000 and $100,000. A few years later a national survey was conducted of the so-called "big money winners," those who had won $50,000 or $100,000 or more, and not a single person had any more money than he did before the big winnings. Their standard of living did not increase. Their life styles were the same as before. They had apparently taken the bonanza and squandered it away with no thought of investing.

One New York bank robber who got away with over a million dollars was found not more than one year later on skid row with nothing to his name. Merely having money does not raise your consciousness about what to do with it. If you were to take all the people on the planet and give them each an equal amount of money, within days you would find some rich and some poor. Take money away from a Rockefeller today and tomorrow he'd be a millionaire. You are not rich because you have money; you have money because you believe you are rich!

On the Job

So you want another job because the one you have now is not paying you enough. Stop and think about it first. Many people switch jobs to get more money, and they do, in fact, get more money. But then they have just that much more to put right back into extra bills.

Their consciousness couldn't accept the extra wealth. What you feel inside is the major factor in determining what you receive for the work you do. To increase your prosperity you must enlarge the inner value you have set. Before anyone else will believe you are worth more, you must be convinced yourself.

The same holds true for getting a raise. If you approach your boss directly with, "I deserve a raise," you will probably get it if you believe that you will. The question is, will you keep it? You must feel you're worth it. I like the old saying, "Do a little more than you get paid for, and it won't be long before you'll be getting paid for a little more than you do."

Doing more than you're paid for is not wasted effort. It has been proven to be an extremely solid rule of life. Understand that sooner or later you are always compensated for your efforts. Even if you don't receive the return from your present position I guarantee you it is like compound interest; the day of collection will arrive. Each day your payoff is building.

Highway Robbery

There is one catch. Just showing up and being warm is not to be confused with working. Collecting a paycheck for work not done could be thought of as stealing time.

I used to be a news reporter, and I had great fun on company time. I'd go swimming, to picnics with my favorite girls, take naps in the park, and so on. I was young, and I wasn't very experienced in life. Ignorance was one of my most noted qualities—ignorance of the laws of life. I could never understand why "fascinating me" was never promoted to office manager. Finally, I quit—out of fear of getting fired. What I didn't realize at the time was that it doesn't take much effort to be a whole lot better.

The best job guarantee that a person can have is the one he gives himself. Doing the job and going one step beyond will guarantee security and a peace of mind that most people strive to achieve all through their lives. The financial benefits are also there for the asking. Wanting something better for yourself is natural, and keeping your eye on the job of the guy ahead of you is okay too, as long as you want a better job for him.

Gloria, a friend of mine, had her eyes on her boss's job, and she told him so. She approached him one day and said, "I want your job and I'm going to get it." Then she added, "That's okay, though, because I want a better one for you." I have found it useful to want for others what you want for yourself. You are a channel for money. You benefit, and so do the people on both ends. Remember, a poor man can be happy but no happy man is poor.

Just for the Fun of It

If you want to switch jobs, just for the fun of it try this. Make a list of a hundred things you like. Really let yourself go—put down anything you like that comes to mind. Out of that list pick out ten that are more important to you than the others — ten that you possibly like better than the other ninety. Now, figure out how you can make money while still being involved with that thing. Sound crazy? It works; try it. You see, the mind works best within limitations. For example, think of something nice. Now think of something nice about your hand. Once you focus your mind on something it will become extremely creative and open within that limited area.

This exercise was done in one of the prosperity seminars and one young man had on his list "dressing well." He liked to dress up, even to the extent of white tie and tails. Once we began to focus into that area all sorts of ideas came to mind. It was suggested that he could be a live mannequin. He could start his own escort service. He might be a model. But the winning suggestion was this: He could sell his presence to restaurants, and that's just what he did. He got $50 an hour for sitting in the dining room of various restaurants looking sophisticated and elegant. He parked his rented Rolls Royce outside and he and an attractive woman sat in the restaurant for an hour or so, eating a free meal. His example shows how it's possible to have some fun with this exercise.

Be a Good Receiver

One way to increase your income is to learn to be a good receiver. For every giver there has to be a receiver. Like anything else this skill must be developed.

For example, many of us feel awkward when someone compliments us on our clothes. Do you respond by saying, "Oh, this old thing?" Or if the subject is your good cooking do you say something like, "It's just something I threw together." By accepting a compliment we allow the other person to give at his or her own level. In a sense we are both givers and receivers. Learning to accept compliments will loosen up the blocks to prosperity in your life and will open you up to be a pipeline to wealth.

Never refuse any good thing offered you. If someone offers you one of the new toasters they received at the wedding shower and you already have one, take it and say thanks. You can find a good home for it somewhere. By not taking it you block their good. They are giving out of love; you can receive in the same manner. If after eating in a fine restaurant your friend wants to pay the bill, allow him. By accepting you are allowing him his good.

How much would you take for your life? Are you worth it? No one can put a figure on his life because life cannot be quantified in terms of money. What about part of your body? Would you take $1,000 for an arm or leg? If someone wanted to buy one of your eyes, how would you feel? Even parts of the body are not up for sale. You are worth the very best in life. You deserve it and you can have that if you feel worthy.

You deserve the very best for many reasons. If you're 30 years old you've been living, playing, having fun, sleeping, eating and digesting all that food for three decades. Your body has been alive for 30 years. You have stored a wealth of information in your brain; that alone should be worth something. You have made friends and family happy, you have contributed to life by your involvement in life, and your very existence itself is reason enough for you to reap the benefits life has to offer. You don't have to do great things to justify your existence; being is its own justificaton.

You will never allow yourself to have more money than you think you are worth! If you want to reap the prosperous life the first thing you have to do is to start feeling good.

The Professionals

Pretend you have some money to invest, even a small amount. Use your imagination. You've never really known what to do with extra money—where to spend it, how to invest it, what to buy. It all seems so complicated. But it's not! It is true that we live in a specialized world. You do what you do well because you trained for it and worked at it to develop the skill necessary.

There are people who, through their strong drives and desires, have developed a skill in handling money. They have insights and an assurance which comes from their experience. These people are the professional money people. They are stock brokers, real estate agents, insurance agents, and bankers. They are the people who have made handling money their business, and their whole life is devoted to helping other people invest their hard-earned money in safe, secure investments which will develop a reasonable return on the dollar.

None of us can become highly experienced in every type of business. If you have a desire to put your money to work effectively and you don't want to take the time to become a stock broker or real estate agent yourself, seek out one of these professionals and have him or her go to work for you, to advise you and give you proper direction for your money. I suggest that your selection of this individual be determined by asking yourself how do you feel about this person, and then proceeding accordingly.

Brown Bag to Riches

A middle aged, mild-mannered man worked in an office of a large corporation on the West Coast. He was a quiet sort who didn't make much fuss about anything. He did his job well: he was punctual and efficient and he earned a reasonably good salary.

For years he joined his friends for lunch. They would go to one of the local restaurants for an hour or so and then return to work. Gradually, however, he began to remain in the office to read during his lunch period, and this finally developed into a standard practice. He would bring his brown bag lunch and when the clock struck noon he'd put his work away, bring out his lunch and begin reading *Barron's*. He read each issue from cover to cover. He had developed an interest in investing.

Although he was not a broker himself, he observed the investments made by several of the larger successful corporations. He followed the trends and then he began buying what they bought and buying when they bought. He also sold what and when they sold.

Time passed, and the day came when he announced to his office crew that he was quitting. Amazement filled the office, for they all wondered what this meek, unassuming man was going to do. He was asked how he could afford to quit, what with the salary and fringe benefits like sick leave and pension, and he said, "I can't afford *not* to quit." When his friends asked him what he meant he answered, "I'm going to be an investment counselor."

"*You?*" was the response. "How can *you* make money at that?"

Then he told them that through his part-time efforts he had already accumulated a quarter of a million dollars. "I just can't afford to keep working here."

This man's interest developed to such a degree that he decided to change jobs. There is no need for you to do that, unless of course you have that driving desire. If your interest is not quite that strong then seek out one of the professionals to assist you in your business transactions.

If you absolutely do not have any money to invest, then it's time for some new word choices. You've got to begin putting yourself on your payroll, and one way to get going is to start saying something like, "A part of all I earn is mine to keep."

So many people pay everyone except themselves — the grocer, doctor, gas company, church. But they forget themselves. That says, "I'm not worth it."

You *are* worth it, and it's time to start putting something aside for *you!* No matter what you make as a salary, you can put something aside for yourself. Some people will say, "But I can't, my expenses exceed my income." I say that they don't know what expenses are. Something is only an expense when you pay for it. If you have any money at all, either in your pocket, in the house, or in the bank, any money at all, then your expenses don't exceed your income. Here's why.

The bills in the desk drawer are not expenses; they are pieces of paper in the desk drawer. They become expenses when you pay them. A telephone call asking you to pay these bills is not an expense; it is someone calling you on the phone. A man knocking at your door wanting payment for these bills is not an expense; it is a man wanting to get in. I repeat, if you have any money at all your expenses do not exceed your income.

There are priorities in life, and you ought to be the highest on your list. Let's face it, if you're not around these bills are not going to get paid anyway. So I suggest that you begin with the attitude of respecting your right of justified financial return. This does *not* mean "me first and to heck with the bills." It *does* mean that you will now allow a portion of your earnings to be put aside for you for the specific purpose of developing a money consciousness, thereby enabling you, in the long run, to better meet your expenses. After all, a change is necessary if things are going to be different. If you continue doing what you're doing, following the same patterns, same thoughts and attitudes, nothing can change. Change is only possible when you change something.

12
The Money Plan

If thou wouldst keep money, save money;
If thou wouldst reap money, sow money.

—Thomas Fuller:
Gnomologia, 1732

Assuming that you have decided that you can live on 90 percent of your income, what are you going to do with the extra 10 percent? You may choose to use the plan that follows for six months to a year to achieve an attitude of abundance. It's great to have mental techniques to use, but it's also nice to have something physical to do to assist you in holding those thoughts. Here is the plan.

Pick a bank or savings firm. Choose one that you really like. You should feel good when you walk into it; it may help to choose one with a particularly attractive design. It smells of money. In fact, all you have to do to get rich is breathe while you're there. Acquaint yourself with the people who work there. It's always a help psychologically to walk into a bank and see the bank president or loan manager, tellers, and even the guard look up and wave or say hello.

Go over to the new accounts section and tell the person you'd like to open up five new savings accounts. Now here's a little test for you. How does this person respond? The response will be directly related to how *you* feel about opening up these accounts. Unsure? Strange? Awkward? Or are you secure and expectant of saving money? The new accounts representative will respond to your feelings. One person told me that she was asked why she was opening up the accounts; was it to hide or shelter some of her income? Her attitude was being reflected right back to her.

When I opened them up I was excited with the idea of building a plan to help myself and others. The initial plan I was exposed to was a Leonard Orr concept. I read it over, made some modifications to suit my personal situation and was waiting for the bank to open on Monday morning. The woman who helped me said, "This looks interesting. What are you doing it for?"

I told her I was developing ideas for a money seminar and when I got it perfected I'd like to conduct one for the bank. I did!

The following are five key savings accounts in developing a money consciousness. This is the suggested outline: you must make your own personal applications.

The Annual Income Account

All your income goes into this account; *all* of it! The reason is that it gives you the experience of having money. It also provides you with an accurate record of your income for your files. You obviously will not be leaving all of your money in this account; you have to use some of it to live on. So, deposit it and at the same time submit a withdrawal slip for dispersement into the other four accounts. Most of the money will be in the account for a very short while, but it is important that you experience depositing the money and seeing it listed.

Of the money deposited you may use nine-tenths for your monthly budgeted needs. Keep the rest in the account. Of all the money you receive and deposit in this account, keep 10 percent in the account and let it build up to the sum of one year's income. If it is not possible for you to keep at least 10 percent in the account from each deposit then keep less, but keep something from each deposit. That is a must! Let it accumulate to equal one year's income.

The savings account may bring only a small return on your money, and you may want to invest in something more profitable. However, the savings account is for the specific purpose of developing a money consciousness, and that's all, so it is important that, to get started, you follow this plan.

While the following example is not recommended for everyone, it does illustrate the importance of the procedure.

A lawyer with an annual income of about $50,000 was in financial trouble. The suggestion was made to deposit one year's income into an annual income account. But one year's income he didn't have. What to do? By liquidating some of his holdings and by borrowing he was able to acquire $50,000, which he then deposited into his annual income account.

Anyone can see that borrowing money at 10 to 15 percent interest and depositing it in a savings account at 5 percent is not going to make any money, and if carried too far it is obviously a losing proposition. However, the purpose of this account is to develop a money consciousness. The lawyer now had one year's income. He could see it sitting there, and as a result he felt that the pressure was off and he could relax a little. Within six months he had developed business interests totaling over $100,000 — more than double his normal annual income. Why? Because he was more at ease about money. He stopped trying so hard to make money, and it began to come. Don't try to make it; you can make a lot of money if you're not trying so hard.

The Financial Independence Account

This account may give you a little trouble at first, but once you get the hang of it you'll see what I'm driving at. *Never* withdraw any money from this account. You may deposit all you like, but once you deposit it you will never see that money again. Kind of scary, right? But there is a very good reason for this account. Because of it, you always have money in the bank. You're never broke. It's a psychological trick and it works.

The question invariably comes up, "What if I have accumulated $10,000? Surely then I should take some of it out?" But consider this. If you have piled up $10,000 in your savings account, you're probably doing pretty well financially. You really don't need the $10,000 to spend; you need a $10,000 idea, which you will get simply because of the fact that you now have a money consciousness. So the rule remains: don't ever spend it.

Obviously you may become so wealthy that you no longer need to play with these savings accounts, but until you reach that point these accounts can't possibly do you any harm; on the contrary, the benefits will far surpass the little extra trouble you will go to with deposits and withdrawals.

Another rule with the financial independence account is that you *must* spend the interest accumulated. Ask the bank to mail you the interest check on this acount, and go out and have lunch on the bank. Your money is working for you all the time, whether you yourself are sleeping, waking, or working. Make sure that you deposit something in this account at least each quarter so that each interest check will be larger than the next. The gradual buildup of funds will provide a feeling of security, and soon you'll be on top of it again. I've been rich and I've been poor, and rich is better.

The Instant Account

This account is the opposite of the financial independence account. In using the instant account you must *spend* all the money you deposit. Maintain a minimum balance and consider that the money deposited is already spent. Deposits in this account are made with the specific purpose of spending randomly—having fun with your money.

For example, you may be out shopping and spot a sweater that you like very much, but the price is $50. Now normally you would walk away fast from a $50 sweater, but these are different times. You're on your way up the ladder to financial success. You check your instant account balance, and find that you have $54.32. The decision is made. You buy the sweater.

Suppose your balance had been only $45.62. Then you wouldn't buy the sweater. Your instant account didn't have enough money. The instant account greatly simplifies the problem of impulse buying; you know immediately just how much money you have to spend on something that comes up on the spur of the moment.

Another technique is helpful, whether you are using the instant account or making cash purchases. Suppose you are considering two shirts. One is higher in price, but you like it just a little better than the other. Then again, you really didn't want to spend that much for a shirt. What to do? Stop for a minute and decide what to you is a nominal amount of money. Is it one dollar? Five dollars? Ten? Let's say that you now have decided that to you five dollars doesn't really make much difference one way or the other. You have now added some flexibility to your buying. If the higher priced shirt is more than five dollars higher than the other, you won't buy it. However, if it is within the five dollar range you can buy it without question. By setting this type of guideline, you relieve yourself of bounding back and forth, and get down to the business of doing instead of wondering.

The Whatever Account

This one you get to name. This one is special because it is devoted to one of your dreams, such as that long yearned for sports car, a boat, or a trip. There is only one time to begin, and that's now. If you don't begin making plans for what you want in life you'll always be "going to," but never quite getting to it. I've heard of all kinds of names for this account: the Hawaii account, the Europe account, the dining room set account, the madness of love account. My own is the helicopter account. I used to fly helicopters and I think they're a kick. I want one, so I'm saving. Another possibility is the free and clear account for clearing up all your past due bills. Notice that it is not called the past due bills account; the emphasis is on the end result, because you want to get the feeling of freedom and completion.

The Investment Account

Deposits made into this account are only withdrawn for the purpose of reinvesting into something which draws a higher yield than your savings account. At present interest rates, alternatives to the savings account are not hard to find. Real estate, the stock market, and bonds are all possibilities, but a word of caution is in order here: beware of "get rich quick" methods. Invest safely and slowly at first; you are building a consciousness of money, not of disaster.

This account may be used for other types of investment than interest-bearing plans. For example, it could be an account for self-improvement programs. An investment in yourself is one of the greatest investments you could possibly make. Education and personal development programs cannot be evaluated in terms of percentage returns, but the value derived can be far greater than the first glance of the eye can see. So many people desire to better their circumstances but are unwilling to better themselves. Oliver Wendell Holmes once said, "The greatest tragedy in the world is not the waste of natural resources, although that is a tragedy. The greatest tragedy in the world is the waste of human resources." An investment in you can be the turning point for that brighter future you've dreamed of, for what lies ahead is directly determined by your zest and enthusiasm for life. You are your only guarantee for a successful future.

Up in Smoke

I know a man who had a fire in his home and his book shelves were burned along with his books—both of them! An investment in a book is the cheapest education you can get. When you are interested in educating yourself, I recommend buying a book rather than borrowing it or going to the library. Build your own library. A collection of books can be extremely valuable for reference, especially the non-fiction, self-help type of books. You can read them over and over, and each time glean something entirely new from them.

The Checking Account

This account is the source of most money problems. There are also more crimes committed through the writing of bad checks than with any other crime in the world.

Writing a bad check is a crime in more ways than one. It's a crime committed against yourself, because it helps you maintain low self-esteem. That's right; the real reason many people bounce checks is that they feel unworthy anyway, so why try. Many people avoid balancing their accounts. Their underlying fear is that they may be overdrawn, and they're scared to find out about it, so they simply avoid the issue. The result is often disaster, extra charges, embarrassment, extra work, and so on. When you get yourself squared away to feeling like the million dollar person that you really are, you'll never bounce a check!

The Format

A checking account is a great convenience when it's handled right. It's good for sending money through the mail, and it provides a record of expenses for tax purposes. But the very convenience is sometimes a trap. Checks are so easy to write that sometimes we misuse the privilege. The solution is to keep an absolute minimum balance in your checking account. Begin a new page in your record book and on the very top line in large letters write DEPOSIT. Don't fill in the amount at this time. Then write checks for the bills you owe, listing them one after another. When you have written checks for all the bills you'll be paying at this time, add up the column of figures. That is the amount your deposit will be, so then enter it on the top of the page.

Then you make out a withdrawal slip from your annual income savings account for that amount. Mail the withdrawal slip and the bills at the same time. The money will be transferred to your checking account by the time your checks are in for payment.

This process insures you that you will never again write a bad check. Also, by keeping a minimum balance you will spend more wisely. Just the act of all that paperwork will keep you alert and in touch with your money.

Here are two other ideas that you may find to your liking. On the top of your checks, during the initial printing, you might like to have printed "Service Bank of the Universe" just above your name, acknowledging your infinite abundance. Another idea that gets people's attention and helps you maintain a positive attitude with money is writing or having printed, "It is a pleasure to Pay to the Order of ..." instead of the usual "Pay to the Order of."

Developing a prosperous attitude takes constant attention, so consider the following idea as well. Tear out about four checks from your check book. On the very top write, "Service Bank of the Universe." Date the check a date that you would like to receive a particular sum of money. Let's say that, by the end of the year, you would like to have earned $20,000. Thus, the date will be December 31, 19____. Next in line write, "It is a pleasure to Pay to the Order of," and here fill in your own name. Also, fill in the amount of $20,000. In the bottom left hand corner indicate the reason for the check: "For excellent services rendered in the United States of America during the year 19____. Then sign it, "The Citizens of America."

Now, take each check and repeat the process. When you have finished post them in conspicuous places around the house—the bathroom mirror, over the bathtub water faucet, on the refrigerator door, or wherever you'll notice them as you go through your day. Remember, frequency and intensity are the keys to creating the visible from the invisible. The ideas on this page are worthless unless you apply them. By using them you will develop a money consciousness.

Cash

Cash can also help to develop that kind of consciousness. Don't be afraid to carry cash with you and use it. During your monthly budgeting, budget a certain amount of cash for your spending out of the annual income account. Actually go to the bank and get yourself a $50 or $100 bill. Carry it with you at all times. Look at it every time you pull your money clip out. (I suggest using a money clip because it is folded and the sum appears larger to you, and you'll have to handle it more often. That's important.)

Don't spend the bill unless you can replace it immediately, but do plan on spending it from time to time just to give yourself a feeling of wealth, even if you only buy a one-dollar item and get ninety-nine dollars in change. This bill is not to flash around; it is for you and your consciousness, and it will provide you with a feeling of excitement. Get rid of the idea of loss or robbery; if by some far remote chance either of these occurs, consider your "loss" seed money and it will return tenfold.

Plastic Money

Charge cards can be extremely useful at times, especially when identification is required. It seems that we exist when we have a plastic card. However, charge cards are a source of anxiety for most people and are definitely a way to avoid payment at the time of purchase. The whole concept is based on scarcity; you don't have money now, but maybe, just maybe, you'll have it later. Charge cards are great for record-keeping and can be very convenient, especially when used correctly, but it is too easy for most people to misuse them, and end in a severe financial bind. Misuse may be one of the reasons you got into a mess in the first place, so if you have charge cards use them with extreme caution.

Getting Free and Clear

At this point it might be helfpful to give some attention to the subject of overdue bills. My experience is that most people are willing to receive any amount per month, just as long as you are making regular payments. Through this they can see that you're making an honest effort to clear up the matter. You don't have to impress people by making a $5,000 payment to clear up the entire balance. They don't want to be impressed, they want to be paid. If you don't have $5,000, but you do have $50, pay that. Make a game out of it. Sometimes it's fun to pay either more or less than the regular monthly payment, and often you'll get a response from the firm you owe. They'll ask, "How come you sent us $23.13 this month when your payment is only $20.00?" You can say something like, "That's the way my accounting is working out this month. Hope you don't mind." They'll think you're a financial genius.

Let's consider the reverse for a moment, and say that someone owes *you*, and it's been a long time since you've received a payment. A courteous phone call can do the trick. Keep in mind that they are as anxious to get you out of their way as you are to get the bill paid. You might suggest a $20 payment on the balance and they might come back with, "I don't have it. I'll send it next month." Now next month to you is 30 days away on the calendar and even farther away to the guy you're talking to, so you come back with, "How about $10 this month?" "No, I don't have that either." "Could you send $5?" "Sorry, but I'm really broke this month." The next question is "Can you send 50 cents?" More than likely the person will say yes, and you say, "I realize that it's only a small amount, but I'm sure you'll make every effort to clear this up as soon as possible."

What you are doing is getting the person in the habit of making regular monthly payments. Habits are contagious. Small payments are better than no payments. Neither side likes to resort to the collection agency, especially the debtor, for at that point he only gets half of the balance due.

I can't recommend strongly enough the need to make regular payments on all of your bills. I believe it is better to make regular payments on all than to pay some off completely and leave others hanging. That can lead to disaster in just a very short while.

Seed Money

This last and most important portion of the prosperity section has to do with a process of making money of which there is absolutely no risk whatsoever. That process is using seed money. Once again, a procedure to increase the money flow in your life.

We're talking about circulation; we're talking about planting seeds. But to be more specific, the process requires action on your part.

Imagine the Sea of Galilee. Flourishing shores, vegetation, water rushing in one end and out the other, down the Jordan River—fish, fowl, growth and beauty. Everything moves rapidly from the Sea of Galilee into the Jordan River, and then on into the Dead Sea. Here, instead of life, there is a stench; no growth, only decay. The water comes in, but there is no exit, no movement, no circulation. Freshness is nowhere to be found.

This illustrates the necessity of circulation in life, for to maintain life, constant movement is essential.

Now carry this idea into the seed money concept and see what happens. First of all, it is important to understand the difference between seed money and tithing. Tithing is offering 10 percent of your already earned income. Seed money is the giving of a certain sum of money with the expectation of a tenfold return.

It should be realized that the most effective use of money is when you are seeding with the expectant return. Offering to assist someone or some organization with money is fine, but keep in mind that your assistance is only temporary. For that organization to grow financially its consciousness must be raised. It matters not how much you give; it will remain bound by the limited thinking of its leaders, and therefore return to the level to which they can accept themselves as being in.

Using the farmer's crop as the example, the farmer can only plant what he has to plant. If he has 100 seeds, that is all he can plant. Now to increase his crop he must plant them, grow them and take the new seeds that were grown, which will be thousands, and then replant to produce a new crop. Increase comes out of what you already have!

To get more of what you have you must plant, or seed, part of that which you have.

What would happen if the farmer took his seeds and threw them all on a hot, black, asphalt parking lot? Probably the seeds would wither up and die, strictly from lack of interest. Then they'd blow away, never to be seen or to reproduce again. You can see, then, that it is extremely important where you plant your seeds, whether you are planting actual corn seed or a money seed. Planting a corn seed deep in rich fertile soil, with plenty of water, attention and caring, would certainly be an excellent choice. In planting money seeds, your belief is the soil of the mind and your expectancy of return is the water that is so necessary for growth.

If your belief is so important it would be wise, then, to properly select just where you feel the seed money would do the most good so it may multiply. Would you take a $100 bill and throw it into a lake? That would be similar to the farmer taking his seeds and throwing them on the asphalt parking lot. There would certainly be better places to plant your seeds to ensure the most abundant growth.

The ideal way to seed your wealth is to give a certain percentage of your money — bonds, stocks, land, property, or any other form of material wealth— for the propagation of the Truth, usually in support of those churches, organizations, or activities which are engaged in the dissemination of that Truth.

John D. Rockefeller, Jr. once said, "I have been brought up to believe, and the conviction only grows upon me, that giving ought to be entered into in just the same careful way as investing." Many times I've heard people say that the powers are limited by specifying a certain amount. This is not necessarily true for all you need to do is to re-apply the principle again and again. To use it concurrently is also effective. *There is absolutely no risk whatsoever in the principle of seed money!*

What, then, is this "tenfold return" stuff? It's easy to count in tens, and much more believable than counting in thousands. For example, suppose we had a 13.4 percent return on our money. What would be the return on $23.56? You'd need a calculator to figure it out. But with tenfold you just add a zero. Another reason is this. If I seeded $50 today, I'd have a pretty hard time expecting a $50,000,000 return by next week. That's my problem of course. Believability is important.

The Plant

The farmer in our example may have all the good intentions in the world of planting that little seed. He may go out and buy a special mixture of earth, expensive fertilizer, a new sprinkler system. He may prepare for planting the finest seeds available and have them sitting on the back porch just waiting for the moment. But until he actually plants them *they will not grow!*

When is your seed planted? You too may have every good intention. You locate an organization you feel worthy, a person or group of people that you feel will benefit and re-plant just as you are doing. You write the check, address the envelope, set it on the desk, stamped and ready to go. But until you mail it, it will do nothing. You see, when you drop it in the box you have released it to begin its own growing process. That is your belief in action. You can't get it back; it's gone, and now the only thing left to do is to *expect* the return.

A few suggestions. Plant your seed in the mail box Cultivate your claim by affirming, "I have received my desired return of (state the amount.)." Above all, begin at a modest level; it must be believable. Does it work? Well, does fire release heat? Will an airplane fly? Will a boat float? Yes, of course it will work. It's a law just like the laws of fire, aerodynamics, and flotation. Chance and luck are removed and only law remains.

Taxi into Position and Hold

There you have it — a practical outline for changing significant areas of your life. Used in part, these suggestions can help you clear up many areas of dissatisfaction; used in their entirety, they can alter your entire life course.

Andrew Carnegie once offered a 25 million dollar suggestion: "Take one good idea and use it!" If you heed his advice, it will be much to your advantage. While I will be the first to affirm that we have available to us limitless power, I will also acknowlege that if we dive, head first, into all areas that we think need attention, disaster may very well result. Therefore, I suggest this: pick one or two good ideas from this book and use them first. Directly apply the ideas you choose into specific areas in your life. In the beginning it might be helpful to pick an area that has not totally fallen apart; your emotional involvement will not be as great, and you will probably have an easier time tackling it.

In using these guidelines, please do yourself a favor. Hold on to the ideas, as applied to your particular situation, for at least one week and observe the changes that take place. When you see the dynamics of *you* in action, there will be no turning back.

Also remember that you will make the changes in your life as only you can make them—your course will more than likely be different from anyone else that you know. Ralph Waldo Emerson, in his essay on self-reliance, said, "There is a time in every man's education when he arrives at the conviction that envy is ignorance; that imitation is suicide; that he must take himself for better or for worse as his portion; that though the wide universe is full of good, no kernel of nourishing corn can come to him but through his toil bestowed on that plot of ground which is given him to till."

Whenever you see that someone else has something, is doing something, or is being something that you desire to have, do, or be, don't envy him. Realize that all he did was put the picture in his mind, and through the natural laws of life, it was made available to him. You, too, can put the picture in your mind and it can be yours, but remember, imitation leads to a dead end. You will accomplish this feat, whatever it may be, in your own unique way. You are the best *you* there can ever be. Trying to be like someone else will only put you second on the list—no one can express life exactly as you can in your own unique way, nor can you express it as someone else. It's just not possible!

Also realize that *you* must do it. No one will be solving your problems for you; you must take control of your ship and fly it. A club, a friend, a spouse, or even this book will not do it for you—you must do it! When you commit yourself to that, the wings of freedom are yours.

Whatever you choose to do, remember that thoughts *do* "transcend the world we know," and that through goal-oriented persistence you can achieve what you set out to do. It is your right to fly!

CAUTION

Wake turbulence from departing 747 ahead.
Cleared for take-off.

Index